BRITISH
POULTRY STANDARDS

BRITISH POULTRY STANDARDS

Complete specifications and judging points of all Standardised Breeds and Varieties of Poultry as compiled by the Specialist Breed Societies and recognised by The Poultry Club of Great Britain

Edited by C. G. May
formerly editor of *Poultry World*

NEWNES–BUTTERWORTHS
LONDON   BOSTON
Sydney–Wellington–Durban–Toronto

THE BUTTERWORTH GROUP

ENGLAND
Butterworth & Co (Publishers) Ltd
London: 88 Kingsway, WC2B 6AB

AUSTRALIA
Butterworths Pty Ltd
Sydney: 586 Pacific Highway, NSW 2067
Also at Melbourne, Brisbane, Adelaide
and Perth

CANADA
Butterworth & Co (Canada) Ltd
Toronto: 2265 Midland Avenue, Scarborough,
Ontario, M1P 4S1

NEW ZEALAND
Butterworths of New Zealand Ltd
Wellington: 26–28 Waring Taylor Street, 1

SOUTH AFRICA
Butterworth & Co (South Africa) (Pty) Ltd
Durban: 152–154 Gale Street

USA
Butterworth (Publishers) Inc
Boston: 161 Ash Street, Reading, Mass. 01867

First published in 1954 by Iliffe Books, an imprint of the
Butterworth Group
Second Edition 1960
Third Edition 1971
   Second Impression 1976 by Newnes-Butterworths

ISBN 0 592 00060 5

Printed in England by Chapel River Press,
Andover, Hants.

CONTENTS

BANTAMS

CONTENTS

CONTENTS

ILLUSTRATIONS

INTRODUCTION

To the commercial poultryman of today, the pure breeds of poultry are of little or no consequence. His outlook penetrates no farther than the hybrid strains of layers and broilers and he either does not know, or has found it convenient to forget, that all of them owe their origin to the Standard pure breeds.

Far-seeing geneticists, on the other hand, think differently. Many of them visualise the time when hybrid strains may have to be re-made and it will be then that there will come a resurgence in the demand for pure breeds, not necessarily those that are of outstanding show quality but certainly those whose blood lines are pure. Even today, with new hybrids coming off the pipe-line, there is a consistent call for pure blood lines.

Standards owe their origin to the demand for uniformity in type and colouration of the various breeds. Without them the shows could never have been born and it was as long ago as in 1865 that The Poultry Club authorised the publication of the first " Standard of Excellence in Exhibition Poultry ", an exceedingly modest predecessor to the larger work which was not to appear until after the turn of the century.

Right from those early days, the Club has remained the guardian of the Standards without necessarily being the body responsible for framing them. This task is normally undertaken by the Specialist Breed Club or by the originator of a new breed or variety. So seriously, however, is this guardianship imposed, and accepted by the Clubs, that until a new variety is admitted to Standard it remains unrecognised by show authorities whose events are staged under the Rules of The Poultry Club of Great Britain.

As a result of the incorporation of the British Bantam Association, The Poultry Club finds itself with added responsibilities, particularly by reason of the fact that at most of the shows bantams now outnumber large fowl. There has also been a considerable increase in the number of specialised bantam shows.

Normal procedure for the admittance of a new breed or variety of an existing breed to Standard involves the submission to The Poultry Club by the originator of that breed of live specimens of more than one generation of the birds to be Standardised, a draft of the proposed Standard and a sworn declaration to the effect that the breed reproduces its like to a remarkably high degree. Through its Standards Committee and, finally, its Council, the Club considers the breed and recommends changes in the submitted Standard where changes are necessary or, in some cases, a complete revision. So searching are these investigations that seldom does it occur that a breed is accepted on first application.

The only deviation from this programme is made when a recognised breed is imported from another country in which it has already been accepted to Standard—the Club usually accepts these without the necessity of submitting specimens as sworn declarations.

This is the third edition of British Poultry Standards. In it are a number of major revisions without excluding those features which claimed wide popularity for the earlier works.

Retained are the coloured sections featuring no fewer than 47 breeds of large fowl and bantams in their natural colours and four plates of feathers copied from specimens taken from prominent show winners. While each of the feathers is identified to its breed (or breeds), the actual colour descriptions of them in the appropriate Breed Standard can also be applied to colour descriptions of other breeds. Mahogany, for instance, is the same in Indian Game as in Gold Laced Wyandottes or Millefleur Belgian bantams; the salmon of the Brown Leghorn is also the salmon of the Faverolle, and no difference exists between the beetle-green of the Ancona and that of the Black Leghorn.

As a consequence, therefore, the feathers serve generally as a complete colour guide to every one of the Breed Standards included in this publication. Accompanying each perfect Standard feather is an imperfect one of a type which most often occurs in the breeds specified.

Supplementing the full Standards of all the recognised breeds previous editions of this work contained, in addition, short descriptions of breeds almost on the verge of extinction but still occasionally seen on the show bench. The same applies to this edition with the important exception that, where a breed has revealed unmistakable signs of revival, it has been taken out of the "Other Breeds" chapter and given more prominence with a full Standard.

Conversely, and where popularity has waned, other breeds have been relegated to the abridged Standard chapters. Where a Standardised breed, or variety of a breed, is now known no longer to exist that breed or variety has been omitted.

Introduction, by the Club, of a Judges Panel has gone a long way towards standardising judging. Under a system of both practical and oral examination judges are graded according to their capabilities. Newcomers are probably satisfied to qualify for a single breed only, coming up for further examination as their experience widens. The "plum" certificate is, of course, one which qualifies its holder as an all-round judge. Very few of these are awarded.

While every care has been taken in compiling the Standards the form of presentation does not necessarily follow that adopted by the Specialist Breed Clubs. Instead, and for ease of reference, a prescribed pattern has been set without in any way departing from the salient points of any of the Breeds concerned. It is this form of layout that has the approval of The Poultry Club and is, therefore, recommended to all Breed Clubs to follow.

To safeguard publication interests The Poultry Club has agreed not to accept or authorise publication of any alterations to existing Standards for a period of two years from the issue of this edition.

LARGE FOWL Plate 1

1. Australorp, Male
2. Dorking, Silver-grey Female
3. Indian Game, Male
4. Jubilee Indian Game, Female
5. Orpington, Buff Male
6. Old English Pheasant Fowl, Gold Female
7. Ixworth, Male

LARGE FOWL Plate 1

LARGE FOWL Plate 2

1. Redcap, Male
2. Sussex, Light Female
3. Old English Game, Black-breasted Red Male
4. Old English Game, Spangled Male
5. Old English Game, Wheaten Female
6. Scots Dumpy, Female
7. Scots Grey, Male

LARGE FOWL Plate 3

1. Brahma, Light Female
2. Cochin, Partridge Male
3. Silkie, White Male
4. Croad Langshan, Female
5. New Hampshire Red, Female
6. Jersey Giant, White Female
7. Plymouth Rock, White Male
8. Rhode Island Red, Female
9. Wyandotte, Silver-laced Female

1
2
3
4
5
6
7
8
9
R.A.Vowles

LARGE FOWL Plate 4

1. Ancona, Male
2. Campine, Silver Male
3. Leghorn, Buff Male
4. Andalusian, Female
5. Leghorn, White Female
6. Hamburgh, Silver Spangled Female
7. Minorca, Black Female
8. Leghorn, Brown Male

LARGE FOWL Plate 4

LARGE FOWL Plate 5

1. Barnevelder, Double-laced Male
2. Bresse, White Female
3. North Holland Blue, Female
4. Houdan, Female
5. Marans, Dark Cuckoo Male
6. Faverolle, Salmon Male
7. Poland, Silver Female
8. Welsummer, Female

LARGE FOWL Plate 5

BANTAMS Plate 6

1. Rosecomb, Black Male
2. Japanese, Black-tailed White Male
3. Barbu d'Uccle, Millefleur Female
4. Frizzle, White Female
5. Sebright, Silver Male
6. Barbu d'Anvers, Blue Male
7. Modern Game, Black-red Male
8. Modern Game, Pile Female

STANDARD FEATHER MARKINGS
Plate 7

1. Hackle feather conforming to Standard as applying to Brown Leghorn and other males of black-red colouring. Note the absence of shaftiness, black fringing and tipping. Actual colour of outer border varies in different breeds between dark orange and pale lemon. In such breeds saddle hackle should conform closely to neck hackle.

1A. Faulty hackle in same breeds. There is considerable shaftiness, the striping runs through and the feather is tipped with black. Striping is also indefinite and fouled with red.

2. Hackle feather conforming to Standard from Partridge Wyandotte male. There is no shaftiness and the striping is very solid and distinct. In Partridge Wyandottes lemon coloured hackles are a desirable exhibition point.

2A. Faulty neck hackle in the same breed. Note that the black striping runs through to tip and is irregular in shape. There is also a distinct black outer fringing to the gold border.

3. Standard hackle feather from male of Gold Laced Wyandotte and similar breeds with rich bay ground colour. Note intensity of centre stripe, absence of shaftiness and freedom from blemish in outer border. Note also soundness of colour in underfluff.

3A. Faulty hackle feather from similar breeds, showing indistinct striping, with foul colour, shaftiness and black running through to tip. Underfluff is a mixture of red and dark grey.

4. Standard hackle feather from male of Light Sussex and similar breeds of Ermine markings, such as Light Brahma, Columbian Wyandotte and Ermine Faverolle. The demand is for solid black centre with clear white border extending to underfluff. Green sheen is an important feature.

4A. Faulty hackle feather from similar breeds, showing black fringing to border, black tipping and shaftiness in quill. Underfluff also lacks distinction.

5. Perfect tri-coloured hackle feather from a Speckled Sussex male. The black striping is solid, with green sheen, and the border is the desired rich mahogany colour, finishing with clean white tip. Note clarity of undercolour.

5A. Faulty Speckled Sussex hackle feather showing almost complete lack of black striping, varying ground colour in border, and indistinct white tipping.

6. Neck hackle conforming to Standard of Andalusian male. The so-called Andalusian blue is a diffusion of black and white, and in male hackles a dark border or lacing surrounds the slate-blue feather. Undercolour is sound and even.

6A. Faulty hackle from same breed. The colour generally is blotchy and lacing is indefinite.

7. Standard neck hackle of a Rhode Island Red male. No attempt has been made to show the ultra-dark red usually seen in show specimens, but the colour seen here conforms with Standard and should be agreeable for exhibition. Note purity of undercolour —a very important point in this breed.

7A. Faulty hackle feather from the same breed, showing uneven ground colour, black tipping and smutty undercolour, which is a very severe defect in a Rhode Island Red.

8. Hackle from Ancona male, conforming closely to Standard. Note clear V-shaped white tipping, complete absence of shaftiness, rich green sheen and solidity of dark underfluff, a particularly strong point in the breed.

8A. Faulty hackle feather from same breed, showing indistinct tipping of greyish-white and faulty undercolour not dark to skin.

9. Hackle feather conforming to Standard from Buff Orpington male, very similar, except for exact shade, to feathers from other buff breeds, such as Cochins and Rocks. Note even colour throughout, absence of shaftiness and sound colour in underfluff, with quill buff to skin.

9A. Faulty hackle feather from similar breed, showing severe shaftiness, uneven ground colour with darker fringe, and impure undercolour.

16

EATHER MARKINGS Plate 7

1. Standard hackle from Barred Plymouth Rock male and similar breeds. Note the points of excellence—barring practically straight across feather, sound contrast in black and blue-white, barring and ground colour in equal widths, and barring carried down underfluff to skin. Tip of feather must be black.

1A. Faulty saddle or neck hackle from similar variety. There is lack of contrast in barring, with dull grey ground colour and V-shaped bars.

2. Hackle as Standard description from Silver Campine, in which males are inclined to hen feathering. Note that the black bar is three times the width of ground colour and tip of feather is silver.

2A. In this faulty hackle (also from Silver Campine male) ground colour is too wide and barring narrow. Feather is without silver tip.

3. Standard hackle from Marans male. In this and some similar breeds evenness of barring is not essential, but it is expected to show reasonable contrast. It should, however, carry through to underfluff.

3A. From the same group of breeds this feather is far too open in barring and lacks uniformity of marking. It is also light in under-colour.

4. Standard markings of female body feather in Plymouth Rocks and similar barred breeds where barring and ground colour are required to be of equal width. Note that barring runs from end to end of feather and that tip is black.

4A. Faulty feather from same group. Note absence of barring to underfluff and V-shaped markings; also blurred and indistinct ground colour.

5. Sound body feather from Silver Campine female showing Standard silver tip and barring three times as wide as ground colour, as in the male. Gold Campine feathers are similar but for difference in ground colour.

5A. Faulty female feather, again from Silver Campine. Here again, as in 2A, barring is too narrow in relation to silver ground colour and tip of feather is black.

6. Body feather from Marans female, conforming to Standard requirements. Note that the markings are less definite than in Rocks and Campines, and the black is lacking in sheen, while ground colour is smoky-white.

6A. Faulty Marans female feather. Lacks definition and contrast in barring, which is indefinite in shape, the blotchy ground colour making an indistinct pattern.

7. Excellent body feather from Partridge Wyandotte female, showing correct ground colour and fine concentric markings. Note complete absence of fringing, shaftiness and similar faults. Fineness of pencilling is a Standard requirement.

7A. From the same breed this faulty female feather shows rusty red ground colour and indistinct pencilling, with faulty underfluff.

8. Body feather of Standard quality from Indian or Cornish Game female. The illustration shows clearly two distinct lacings with a third inner marking. Lacing should have green sheen on a rich bay or mahogany ground.

8A. Faulty feather from same breed. Missing are evenness of lacing and central marking. The outer lacing runs off into a spangle tip.

9. Standard feather from Laced Barnevelder female. In this breed ground colour should be rich with two even and distinct concentric lacings. Quill of feather should be mahogany colour to skin.

9A. Faulty Barnevelder female feather, showing spangle tip to outer lacing and irregular inner markings on ground colour that is too pale.

EATHER MARKINGS Plate 8

STANDARD FEATHER MARKINGS
Plate 9

1. Standard markings on Silver Laced Wyandotte female feather, showing very even lacing on clear silver ground colour and rich colour in underfluff. In this breed clarity of lacing is of greater importance than fineness of width.

1A. Faulty female feather from same breed. In this there is a fringing of silver outside the black lacing, which is irregular in width and runs narrow at sides. Undercolour is also defective.

2. Excellent feather from Gold Laced Wyandotte. In this the ground colour is a clear rich golden bay and there is a complete absence of pale shaft. Undercolour is sound and lacing just about the widest advisable.

2A. This shows a very faulty feather from same breed. It portrays mossy ground colour with blotchy markings and uneven width of lacing at sides of feather. Undercolour is not rich enough.

3. Standard markings on Andalusian female feather showing well-defined lacing on clear slate-blue ground and good depth of colour in underfluff. The dark shaft is desirable and is not classed as a fault.

3A. Faulty feather from female of same breed. In this the ground colour is blurred and indistinct, and the lacing is not crisp, while undercolour lacks depth.

4. This shows a feather from an Ancona female, almost perfect in Standard requirements. The white tipping is clear and V-shaped and undercolour is dark to skin.

4A. Faulty feather from female of same breed. Here the tip of feather is greyish-white and lacks the necessary V-shape, while undercolour is not rich enough.

5. An almost perfectly marked feather from a Speckled Sussex female—though the white tip might be criticised by some breeders as rather too large. The black dividing bar shows good green sheen and the ground colour is rich and even.

5A. As a contrast this faulty feather shows a blotchy white tip and lack of colour in underfluff. The ground colour is also uneven.

6. An excellent example of " mooning " on the feather of a Silver Spangled Hamburgh female. Note the round spangle and the clear silver ground with sound undercolour.

6A. In this feather from the same breed the spangling at tip is not moon-shaped and there is too much underfluff and insufficient silver ground colour to body of feather.

7. A good example of the desired colour in Rhode Island Red female plumage. Note the great depth of rich colour and the sound dark undercolour.

7A. Faulty colour in a feather from the same breed. Here the middle of feather is paler and inclined to shaftiness, and colour generally is uneven.

8. Standard plumage in females of Australorp and similar breeds of soft feather with rich green sheen. Note the brilliance of colour and general soundness of underfluff.

8A. This shows a common fault in similar breeds, a sooty or dead black colour without sheen and lacking lustre. This sootiness is, however, usually accompanied by dark undercolour.

9. Standard colour and feather in the Buff Rock female and similar breeds which perhaps vary in exact shade and in quantity and softness of underfluff. Note the clear even buff and lack of shaftiness or lacing, also the sound rich undercolour.

9A. This feather from a similar buff breed shows very bad faults—mealiness and bad undercolour with a certain amount of pale colour in shaft.

EATHER MARKINGS Plate 9

STANDARD FEATHER MARKINGS
Plate 10

1. This shows a typical Standard-bred feather from a Derbyshire Redcap female. Note the rich ground colour and the crescentic black markings, which are really midway between spangling and lacing.

1A. In this faulty feather from a female of the same breed the ground colour is uneven and lacks richness, while the black tip is too small and indefinite and too closely resembles moon-shaped spangling.

2. This is a Standard example of the webless type of plumage associated with Silkies in which the feather vane has no strength and the barbs no cohesion. This plumage is common in all colours.

2A. Faulty feather from the same breed. In this the middle of feather is too solid and lacks silkiness, while the fluff has insufficient length.

3. A delicately pencilled body feather from a Silver-grey Dorking female. Note the silvery colour and absence of ruddy or yellow tinge in ground colour. This type of feather is also usual in Duckwing females of various breeds.

3A. Faulty colour in female feather from same breed. Here there is a distinctly incorrect ground colour and pronounced shaftiness.

4. A good example of Standard-bred colour and markings in body feather of Brown Leghorn female, where the ground colour is a soft brown shade and the markings finely pencilled. This type of feather is common to many varieties of Partridge or Grouse colouring.

4A. This shows a body feather from the same breed, in which ground colour is ruddy and shaftiness is pronounced—both severe exhibition faults.

5. A well chosen example of the irregularity in markings of an Exchequer Leghorn female. In this breed the black and white should be well distributed but not regularly placed, and under-fluff should be parti-coloured black and white.

5A. This faulty feather from the same breed shows a too regular disposition of markings, the body of the feather being almost entirely black and the white markings almost resembling lacing.

6. This is a Standard feather from the breast of a Silver Dorking, and with slight variations of shade from pale to rich salmon applies to a number of varieties with Black-red or Duckwing colouring. Colour should be even with as little pale shaft as possible.

6A. A faulty sample of breast feather from the same group. Here the ground colour is washy and disfigured by pale markings known as mealiness.

7. Standard markings in North Holland Blue female. Note the defined but somewhat irregular barring on a distinctly bluish ground. No barring or other requirements in underfluff are called for in the Standard.

7A. This shows a faulty female feather in the same breed, which is not closely Standardised for markings. The ground colour is smoke-grey instead of blue, and is blotchy, with uneven markings.

8. A good example of clear colour in an unlaced or self-blue female feather, where no lacing is permissible, such as in Blue Leghorns, Blue Wyandottes, etc. Note even pale-blue shade and absence of any form of markings. This is an example of the true-breeding blue colour found in Belgian Bantams.

8A. This faulty female feather is a dull dirty grey instead of clear blue, and has blotchy markings as well as a suggestion of irregular lacing.

9. A good sample of exquisitely patterned thigh fluff in Rouen drakes. The ground colour is a clear silver and the markings a delicate but clear black or dark brown. These markings are sometimes known as chain mail.

9A. Another good Rouen feather—this time from the duck. Ground colour is very rich and markings intensely black, though seldom so regular and even as in domestic fowl.

1
1A
2
2A
3
3A
4
4A
5
5A
6
6A
7
7A
8
8A
9
9A

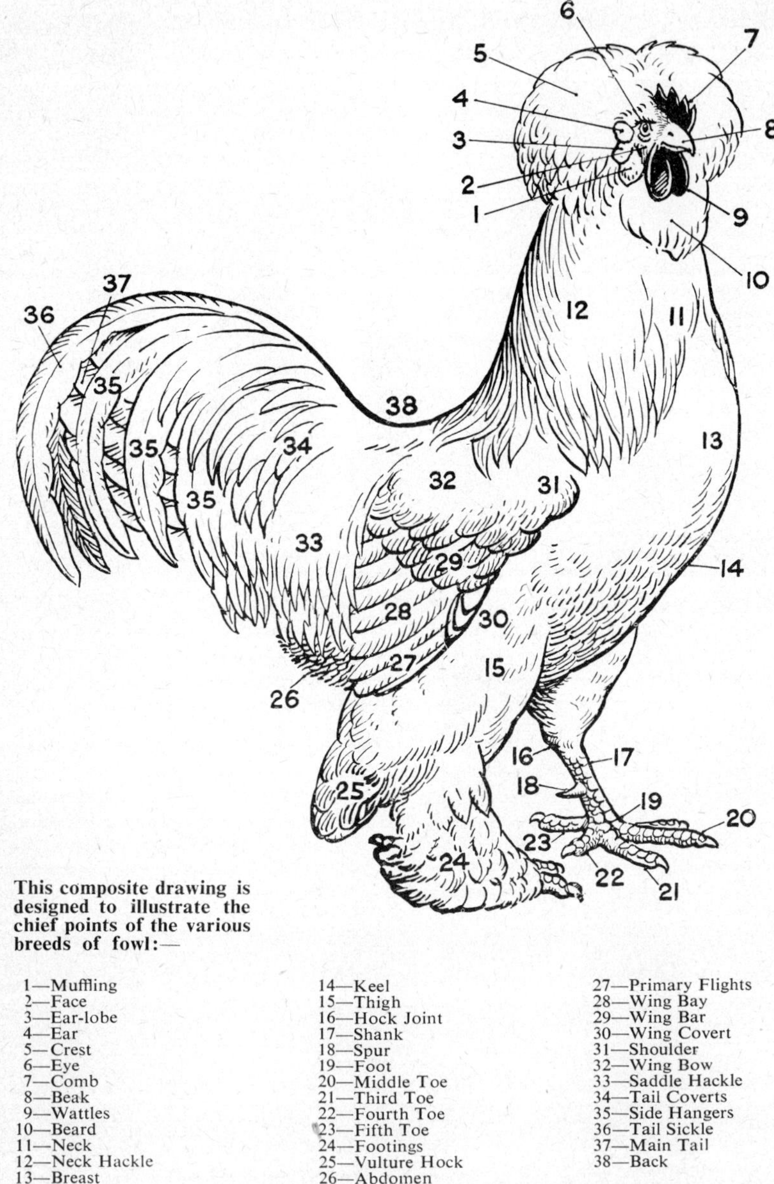

This composite drawing is designed to illustrate the chief points of the various breeds of fowl:—

1—Muffling
2—Face
3—Ear-lobe
4—Ear
5—Crest
6—Eye
7—Comb
8—Beak
9—Wattles
10—Beard
11—Neck
12—Neck Hackle
13—Breast

14—Keel
15—Thigh
16—Hock Joint
17—Shank
18—Spur
19—Foot
20—Middle Toe
21—Third Toe
22—Fourth Toe
23—Fifth Toe
24—Footings
25—Vulture Hock
26—Abdomen

27—Primary Flights
28—Wing Bay
29—Wing Bar
30—Wing Covert
31—Shoulder
32—Wing Bow
33—Saddle Hackle
34—Tail Coverts
35—Side Hangers
36—Tail Sickle
37—Main Tail
38—Back

ANCONA

ORIGIN: Mediterranean
CLASSIFICATION: Light
EGG COLOUR: White to Cream

NAMED after the province of Ancona in Italy, specimens of this Mediterranean breed were imported into England in 1851, first the single then the rose comb. Controversy centres around the view that Anconas are akin to the original Mottled Leghorn and, therefore, a member of the Leghorn family. However, the fact remains that breeders adhere to the name of Ancona. The breed has retained its popularity on the show bench not only for its laying propensities, but because of its combination of breed type and characteristics with usefulness.

GENERAL CHARACTERISTICS: MALE

Carriage: Alert, bold and active.

Type: Body broad, close and compact. Back of moderate length. Breast full and broad, carried well forward and upward. Wings large and carried well tucked up. Tail full and carried well out.

Head: Deep, moderate in length, rather inclined to width, and carried well back. Beak medium with a moderate curve. Eyes bright and prominent. Comb single or rose. The single of medium size, upright and with five to seven deep broad even serrations forming a regular curve, coming well back and following the line of the head, free from excrescences. The rose resembles that of the Wyandotte. Face smooth and of fine texture. Ear-lobes medium, inclined to almond shape, free from folds. Wattles long, fine in texture, in proportion to comb.

Neck: Long, nicely arched and well covered with hackle.

Legs and Feet: Legs of medium length, strong, set well apart, clear of feathers, thighs not much seen. Toes, four, rather long and thin and well spread out.

FEMALE

With the exception of the single comb, which falls gracefully to one side of the face, without obscuring the vision, the characteristics are generally similar to those of the male, allowing for the natural sexual differences. The body, however, is round and compact, with greater posterior development than in the male. The back is rather long and broad, the neck of medium length and carried well up.

COLOUR

Plumage, Male and Female: Good beetle-green ground tipped with white (the more V-shaped the better). No inclination to lacing. The more evenly V-tipped throughout with beetle-green and white the better, provided the ground colour is beetle-green.

In both sexes: Beak yellow with black or horn shadings; a wholly yellow beak not desirable. Eyes, iris orange-red, pupil hazel. Comb, face and wattles, bright red, face free from white. Ear-lobes, white. Legs, yellow, mottled with black, the more evenly mottled the better.

Standard Weights: Cock 6 to 6½ lb; cockerel 5½ lb. Hen 5 to 5½ lb; pullet 4½ lb.

SCALE OF POINTS

TYPE AND CARRIAGE	15
TEXTURE (GENERAL)	10
SIZE	5
PURITY OF WHITE, QUALITY AND EVENNESS OF TIPPING	20
BEETLE-GREEN GROUND COLOUR DARK TO SKIN	15
LEG COLOUR	5
HEAD (EYE 5, COMB 10, LOBE 5)	20
BEAK COLOUR	5
CONDITION	5
	100

(*Note:* Eye points include brightness and prominence. Comb points include medium size and fine texture.)

DEFECTS

	To lose
IN-KNEED	10
SQUIRREL TAIL	10

Continued on page 28

26

Ancona
Male (single comb)
Female (rose comb)

Ancona defects—continued

CROOKED TOES	10
WHITE OR LIGHT UNDERCOLOUR	10
GROUND COLOUR OTHER THAN BEETLE-GREEN	10
TAIL NOT TIPPED OR NOT BLACK TO ROOTS	10
WINGS ANY OTHER COLOUR THAN BLACK TIPPED WITH WHITE	10
BAD COMB	5
WHITE IN FACE	20
LOBE OTHER THAN WHITE	5
	100

(*Note:* Roach back or any bad structural deformity a disqualification.)

ANDALUSIAN

ORIGIN: Mediterranean
CLASSIFICATION: Light
EGG COLOUR: White

THE breed owes its name to the Province of Andalusia, in Spain, and is one of the oldest of the Mediterranean breeds. It is a contemporary of the Black Spanish with which, no doubt, it is closely related. The Blue Andalusian, as we know it today, was developed from black and white stock imported from Andalusia about 1846, and blending of the two colours most probably created the Blue. The earlier specimens were large, and game-like in carriage, with medium combs and lobes, and of a self-colour, although individual birds were selectively bred for lacing, by infusion of Black Minorca blood.

GENERAL CHARACTERISTICS: MALE

Carriage: Upright, bold and active.

Type: Body long, broad at the shoulders, and tapering to the tail, with the plumage close and compact. Breast full and round. Wings long, well tucked-up and the ends covered by the saddle

Andalusian
Male and Female

hackles. Tail large and flowing, carried moderately high, but not approaching " squirrel " or fan shape.

Head: Moderately long, deep and inclined to width. Beak stout and of medium length. Eyes prominent. Comb single, upright and of medium size, deeply serrated with spikes broad at the base, the back portion slightly following the line of the head but not touching the neck. Free from " thumb marks " or side spikes. Face smooth. Ear-lobes almond in shape, medium size, free from wrinkles, and fitting closely to the face. Wattles fine and long.

Neck: Long and well covered with hackle feathers.

Legs and Feet: Legs long. Shanks and feet free from feathers. Toes, four, straight and well spread.

FEMALE

With the exception of the comb, which falls with a single fold to one side without covering the eye, the general characteristics are similar to those of the male, allowing for the natural sexual differences.

COLOUR

Plumage, Male and Female: Clear blue, edged with distinct black lacing, not too narrow, on each feather, excepting the male's sickles, which are dark (or even black), and his hackles, which are black with a rich gloss, while the female's neck hackle is a rich lustrous black, showing broad lacing on the tips of the feathers at the base of the neck. Undercolour to tone with surface colour.

In both sexes: Beak dark slate or horn. Eyes dark red or red-brown. Comb, face and wattles bright red. Ear-lobes white. Legs and feet dark slate or black.

Standard Weights: Male 7 to 8 lb. Female 5 to 6 lb.

SCALE OF POINTS

GROUND COLOUR	30
LACING	20
HEAD (COMB 10, FACE 10, LOBES 5)	25
SIZE, TYPE, CARRIAGE, TAIL AND CONDITION	25
	100

Serious Defects: In the male, much white in face or presence of red in lobes. White feathers. Sooty ground colour. Red or yellow in hackles. Any deformity and comb not upright. In the female any of these points that apply, together with an upright comb.

ASEEL

ORIGIN: Asiatic
CLASSIFICATION: Light
EGG COLOUR: Tinted

THE Aseel is one of the oldest of the game fowls of Asia, and is considered to be one of the principal ancestors of the breed originated in Cornwall but known as " Indian " Game.   Never numerous in Britain, the Aseel has been reintroduced from time to time, and occasionally appears at shows.   One of their characteristics is their in-born tendency to fight among themselves from a very early age, often with fatal results, even among youngsters of the same brood.

GENERAL CHARACTERISTICS : MALE

Carriage: Upright, but not too " gamey " in general appearance.

Head: Skull short, broad between the eye and jaw, thick at the base. Beak strong, thick, and somewhat short.   Eyes bold, set back in the head.   Comb triple (or pea) very hard fleshed, and small.   Face hard and fine.   Ear-lobes small.   Wattles none.

Neck: Medium length, the same width throughout, powerful, with scanty feathering and clean throat, not prominent or fleshy.

Body: Short and round, with very close plumage.   Straight back. Broad and high shoulders.   Fairly narrow stern, but thick and strong at the root of the tail.   Short strong wings, level, carried well out from the shoulders. Short tail slightly drooping, sickles tapering like a scimitar to within three or four inches of the ground.

Legs: Thick, muscular and short.   Thighs well apart.   Shanks and feet free from feathers.   Toes (four) straight, and the hind toe, if duckfooted, not a disqualification.

Plumage: Short and wiry, difficult to break, and devoid of fluff. Almost naked at point of breast-bone and first joint of the wings.

Handling: Firm and muscular, very heavy in comparison with size.

FEMALE

The general characteristics are similar to those of the male, allowing for the natural sexual differences.

COLOUR

Except that the comb, face, jaws and throat are red, there is no standard or fixed colour of beak, eyes, legs and plumage. The principal varieties of the breed are Black, Grey, Red, Black Spangle, Red Spangle, White and Yellow.

Standard Weight: Male 6 lb.   Female 5 lb.

SCALE OF POINTS

HEAD (SKULL 20, COMB AND EYES 5)....	25
TYPE......................................	20
CONDITION	20
NECK, STERN, AND PLUMAGE, 5 EACH ..	15
LEGS AND FEET..........................	10
CARRIAGE OF TAIL	10
	100

Serious defects: Wry tail, roach back, or any other deformity.

AUSTRALORP

ORIGIN: **British**
CLASSIFICATION: **Heavy**
EGG COLOUR: **Tinted to brown**

THE claim that the Australorp—an abbreviation of Australian Black Orpington—is the prototype of the Black Orpington, as originally made by Mr. Wm. Cook, has never been questioned. Its breeders emphasise that its true utility type gives to poultrymen the Orpington at its best, an excellent layer and a good table fowl, with white skin.

It was around 1921 that large importations of stock birds were made from Australia into this country and an Austral Orpington Club founded. Later the breed name of Australorp was adopted, and this remains today.

GENERAL CHARACTERISTICS: MALE

Carriage: Erect and graceful, denoting an active fowl, the head being carried well above the tail line.

Australorp
Male and Female

Type: Body deep and broad, showing somewhat greater length than depth. Back broad across shoulders and the saddle, with a sweeping curve from neck to tail. Breast full and rounded, carried well forward without bulging; breast bone long and straight. Wings compact and carried closely in, the ends being covered by the saddle hackles. Tail full and compact, rising gradually from the saddle in an unbroken line; the sickles gracefully curved, but not long and streaming.

Head: Finely modelled with skull rounded. Beak slightly curved, strong, of medium length. Eyes, large, prominent and expressive; high in skull standing out well when viewed from front or back. Comb single, medium in size, erect, evenly serrated (four to six serrations) and blade tending downwards without touching the neck, texture fine, but not of polished appearance. Face full, fine in texture, clean, free from feathers, wrinkles and overhanging brows. Ear-lobes small and elongated. Wattles medium in size, rounded at bottom and corresponding in texture to comb.

Neck: Fairly long, fine at the junction of head, with a gradual outward curve to the back, widening distinctly at the shoulders.

Legs and Feet: Legs medium in length, strong, rounded in front and spaced well apart, the hocks nearly covered by body feathering, and the whole of the shanks showing below the underline. Shanks and feet (four toes) free from feathers or down.

Plumage: Feathering soft but close, with a minimum of fluff, the lower body fluff being only sufficient to cover the thighs.

Skin: Fine in texture.

FEMALE

The general characteristics are similar to those of the male, allowing for the natural sexual differences. The pelvic bones should be pliable, not showing an excess of fat or gristle; the abdominal skin being pliable without an excess of internal fat. All these parts to be of fine texture; any indication of coarseness should be discountenanced.

COLOUR

Plumage, Male and Female: Black with lustrous green sheen.

In both sexes: Beak black. Eyes black or dark brown iris, black preferred. Face, comb, ear-lobes and wattles bright red. Legs and feet black with white soles. Skin white.

Standard Weights: Cock $8\frac{1}{2}$ to 10 lb; cockerel $7\frac{1}{2}$ to 9 lb. Hen $6\frac{1}{2}$ to 8 lb; pullet $5\frac{1}{2}$ to 7 lb.

SCALE OF POINTS

TYPE...................................... 35
HEAD (EYES 10, FACE 5, SKULL 5, COMB
    AND WATTLES, 5)...................... 25
PLUMAGE (COLOUR, QUALITY AND CHAR-
    ACTER OF FEATHERING)................ 12
TEXTURE AND FREEDOM FROM COARSE-
    NESS.................................. 15
CONDITION................................ 8
LEGS AND FEET........................... 5
                                                 100

Defects (for which birds should be passed): Any deformity such as wry tail, roach back, crooked breast bone, crooked toes, webbed feet. Yellow or willow colour in legs or feet. Yellow or pearl coloured eyes. Feathering on shanks or feet. Side sprigs on comb. Split or twisted wing and slipped wing.

Serious Defects: Red, yellow or white in feathers; permanent white in ear-lobes.

BARNEVELDER

ORIGIN: Dutch
CLASSIFICATION: Heavy
EGG COLOUR: Brown

THIS breed was originated in the district of Barneveld, Holland, and stock were imported into this country about 1921, with the brown egg as one of the chief attractions. At first the birds were very mixed for markings, some being double laced, others single, while the majority followed a partridge or " stippled " pattern. Two varieties were standardised, namely, Double Laced and Partridge or " stippled ", but the former gradually came to the top, and is the popular variety of today.

GENERAL CHARACTERISTICS: MALE

Carriage: Alert, upright, and well-balanced, the body appearing compressed, and the back concave.

Type: Body of medium length, deep and broad, with broad shoulders and high-set saddle. Breast and rump deep, broad and full. Wings rather short and carried high. Tail full, with graceful and uniform sweep.

Head: Carried high with neat skull. Beak short and full. Eyes very bold, bright and prominent. Comb single, upright, of medium size and well serrated, with a firm base, the heel to follow the neck. Face smooth, and as free from feathers as possible. Ear-lobes long. Wattles of medium size.

Neck: Fairly long, full and carried erect.

Legs and Feet: Thighs and shanks of medium length to give symmetry. Shanks and feet free from feathers. Toes four, well spread.

Plumage: Fairly tight and of nice texture.

FEMALE

The general characteristics are similar to those of the male, allowing for the natural sexual differences.

COLOUR

Plumage, Black Variety, Male and Female: Black with beetle-green sheen. This variety is now comparatively extinct.

Plumage, Double Laced Variety, Male: Neck and saddle hackles to match for colour and definition, each feather to be black (beetle-green) with slight red-brown edging and red-brown centre quill (stem) finishing black to tip. Breast red-brown with black (beetle-green) outer edging or lacing. Back and cape red-brown feathers with very wide black lacing. Abdomen and thighs black (beetle-green) with black down. Wing, bow and bar, red-brown with broad lacing; primaries, inner edge black, outer red-brown; secondaries, inner edge black, outer red-brown finely laced with black, showing when closed as a red-brown bay. Tail, all main feathers black, with beetle-green sickles and hangers. All visible black feathers and lacing to show beetle-green sheen. Undercolour dark slate.

Plumage, Female: Hackle black with beetle-green sheen. Breast, saddle, back and thighs red-brown ground clear of peppering, each feather with defined glossy black outer lacing, and inner defined lacing, the outer lacing to be distinct yet not so heavy as to give a

Barnevelder
Male and Female

black appearance to the bird in the show pen. Abdomen black with black down preferred. Wing, primaries inner edge black, outer brown; when wing is closed a brown bay is formed; secondaries inner edge black, outer brown, finely laced with black. Tail, main feathers black with laced feathers well up to them; undercolour, grey.

Plumage, Partridge Variety, Male: Neck and saddle hackles red-brown with distinct but small black tip; fluff, grey; quill, red-brown. Breast, black (beetle-green). Abdomen and thighs, black (beetle-green) with black down. Back, cape and wing bow, red-brown with wide black tip; fluff, grey; quill, red-brown. Wing, bar, black; bay, brown; primaries, inner edge black, outer brown; secondaries, inner edge black, outer brown (seen as wing is closed). Tail, main feathers black with beetle-green sheen; coverts, upper black, lower red-brown peppered with black; sickles black with beetle-green sheen. All visible black feathers with beetle-green sheen.

Plumage, Female: Hackle black with beetle-green sheen. Breast, saddle, back and thighs red-brown ground evenly stippled with small black peppering, clear of defined inner lacing or pencilling, each feather with glossy black outer lacing, not so broad as to make the bird appear black when seen in the show pen. Wing, primaries inner edge black, outer brown peppered with black; secondaries, outer edge brown evenly stippled with small black peppering. Tail, main feathers black, coverts peppered; undercolour, grey.

Pumlage, Silver Variety, Male: Hackle silver with black centres. Breast silver with black edging. Back and saddle black centre with white edges. Undercolour silver grey. Wing, primaries black; secondaries black edged with white. Tail, black with beetle-green sheen; sickles edged with white.

Plumage, Female: Hackle black centre with white edges, a little rust permissible. Breast white, slightly peppered, outside edge black. Wing, primaries black inside, white outside, slightly peppered; secondaries well peppered.

In both sexes: Beak yellow with dark point (in the Silver, horn). Eyes orange. Comb, face, wattles and ear-lobes red. Legs and feet yellow.

Standard Weights: Cock 7 to 8 lb; cockerel 6 to 7 lb. Hen 6 to 7 lb; pullet 5 to 6 lb.

SCALE OF POINTS

TYPE AND SIZE......................... 30
COLOUR............................... 25
TEXTURE.............................. 15
HEAD................................. 10
LEGS AND FEET........................ 10
HEALTH AND CONDITION................. 10
                                      ———
                                      100

Serious Defects: White in lobes. Squirrel or wry tail. Feathered legs or toes. Side sprigs on comb. Crooked toes. High or roached back. Seriously deformed breast bones. More than four toes on either foot. Black legs. **Minor Defects:** White in under-colour, flights, tails, wings, sickles or fluff.

BRAHMA

ORIGIN: Asiatic
CLASSIFICATION: Heavy
EGG COLOUR: Tinted

THE origin of the Brahma is wrapped in a deal of mystery, although it is accepted as of Asiatic ancestry. Its original name of Brahma-Pootra, after the river in India of that name, supports the first arrival of such stock in New York in 1846 from India. Stock reached this country in 1853. Much of the confusion at the time may have been associated with the lack of uniformity and of breed characteristics in the extra large and profusely feathered breeds being shipped from China. Both Light and Pencilled Brahmas were included in The Poultry Club's first Book of Standards issued in 1865, and the breed was developed with the pea comb as a characteristic.

GENERAL CHARACTERISTICS: MALE

Carriage: Sedate, but fairly active.

Type: Body broad, square and deep. Back short, either flat or slightly hollow between the shoulders, the saddle rising half-way between the hackle and the tail until it reaches the tail coverts.

Breast full, with horizontal keel. Wings medium sized with lower line horizontal, free from twisted or slipped feathers, well tucked under the saddle feathers, which should be of ample length. Tail of medium length, rising from the line of the saddle and carried nearly upright, the quill feathers well spread, the coverts broad and abundant, well curved, and almost covering the quill feathers.

Head: Small, rather short, of medium breadth, and with slight prominence over the eyes. Beak short and strong. Eyes large, prominent. Comb triple or " pea ", small, closely fitting and drooping behind. Face smooth, free from feathers or hairs. Ear-lobes long and fine, free from feathers. Wattles small, fine and rounded, free from feathers.

Neck: Long, covered with hackle feathers that reach well down to the shoulders, a depression being apparent at the back between the head feathers and the upper hackle.

Legs and Feet: Legs moderately long, powerful, well apart, and feathered. Thighs large and covered in front by the lower breast feathers. Fluff soft, abundant, covering the hind parts, and standing out behind the thighs. Hocks amply covered with soft rounded feathers, or with quill feathers provided they are accompanied with proportionately heavy shank and foot feathering. Shank feather profuse, standing well out from legs and toes, extending under the hock feathers and to the extremity of the middle and outer toes, profuse leg and foot feather without vulture hock being desirable. Toes, four, straight and spreading.

Plumage: Profuse, but hard and close compared with the Cochin.

FEMALE

With the exception of the neck and legs, which are rather short, the general characteristics are similar to those of the male, allowing for the natural sexual differences.

COLOUR

Plumage, Dark Variety, Male: Head silver-white. Neck and saddle hackles silver-white, with a sharp stripe of brilliant black in the centre of each feather tapering to a point near its extremity and free from white shaft. Breast, underpart of body, thighs and fluff, intense glossy black. Back silver-white, except between the shoulders where the feathers are glossy black laced with white. Wing-bows silver-white; primaries black, mixed with occasional feathers having a narrow white outside edge; secondaries, part of outer web (forming " bay ") white, remainder (" butt ") black;

Brahma
Dark Male,
Light Female

coverts glossy black, forming a distinct bar across the wing when folded. Tail black, or coverts laced (edged) with white. Leg-feathers black, or slightly mixed with white.

Plumage, Female: Head silver-white or striped with black or grey. Neck hackle similar to that of the male, or pencilled centres. Tail black, or edged with grey, or pencilled. Remainder any shade of clear grey finely pencilled with black or a darker shade of grey than the ground colour, following the outline of each feather, sharply defined, uniform, and numerous.

Plumage, Light Variety, Male: Head and neck hackle as in the Dark variety. Saddle white preferably, but white slightly striped with black in birds having very dark neck hackles. Wings, primaries black or edged with white; secondaries white outside and black on part of inside web. Tail black, or edged with white. Remainder clear white, with white, blue-white, or slate undercolour, not visible when the feathers are undisturbed. Black-and-white admissible in shank and toe feathering.

Plumage, Female: Neck hackle silver-white striped with black (dense at the lower part of the hackle), the black centre of each feather entirely surrounded by a white margin. In other respects the colour of the female is similar to that of the male.

Plumage, White Variety, Male and Female: Pure white throughout.

In both sexes: Beak yellow or yellow and black. Eyes orange-red. Comb, face, ear-lobes and wattles bright red. Legs and feet orange-yellow or yellow.

Standard Weights: Male 10 to 12 lb. Female 7 to 9 lb.

SCALE OF POINTS

TYPE, SIZE, FEATHER AND CARRIAGE.....	35
HEAD, LEGS AND FEET....................	20
COLOUR AND MARKINGS.................	40
CONDITION...............................	5
	100

Serious Defects: Comb other than " pea " type. Badly twisted hackle or wing feathers. Total absence of leg feather. Great want of size in adults. Total want of condition. White legs. Any deformity. Buff on any part of the plumage of Light. Much red or yellow in the plumage, or much white in the tail of Dark males. Utter want of pencilling, or patches of brown or red in the plumage of Dark females.

BRESSE

ORIGIN: French
CLASSIFICATION: Light
EGG COLOUR: White

THE Bresse, deriving its name from the territory south of Burgundy, is a fairly firmly established favourite in France, renowned for its table qualities.   There have been many attempts to popularise the breed in this country, and it is strange that a bird of such potentialities should fail to make its mark; for it is not to be despised as a layer, and it is quick maturing and hardy.   As to its failure as a table bird in Britain possibly an answer is to be found in the fact that it possesses shanks of a dark slate colour, and on this side of the Channel the prejudice against skin and legs of any colour other than white is extremely strong.

GENERAL CHARACTERISTICS: MALE

Carriage: Active and graceful.

Type: Body fairly broad and compact.   Back moderately long, broad shoulders and saddle.   Breast well rounded and deep. Wings long and carried close to body.   Tail well developed and carried at an angle of 45 degrees with the back; well rounded sickle feathers.

Head: Medium sized.   Beak strong and fairly short.   Eyes bold. Comb single, erect and of medium size, evenly serrated, fine texture, the back part of it (the heel) clear of the neck but following the curve of the head, free from " thumb marks " and side spikes. Face smooth and free of feathers.   Ear-lobes well developed. Wattles of medium length, fine in quality and rounded at ends.

Neck: Of medium length, amply furnished with hackle feathers.

Legs and Feet: Legs moderately long and well apart.   Shanks free from feathers.   Toes, four, straight and well spread.

FEMALE

With the exception of the comb, which falls gracefully over eitner side of the face, the general characteristics are similar to those of the male, allowing for the natural sexual differences.

COLOUR

Plumage, Black Variety, Male and Female: The plumage is black

Bresse
Black Male,
White Female

with a brilliant beetle-green sheen, and the undercolour black.

In both sexes: Beak dark horn. Eyes black or dark brown. Comb, face and wattles bright red. Ear-lobes snow white. Legs and feet blue-grey.

Plumage, White Variety, Male and Female: Pure white, straw tinge objectionable.

In both sexes: Beak blue-white. Eyes black or dark brown. Comb and wattles bright red. Face red or sooty. Ear-lobes blue-white or white (a little red allowed). Legs and feet slate-blue.

Standard Weights: Cock 5½ to 6 lb; cockerel 5 to 5½ lb. Hen 4½ to 5 lb; pullet 4 to 4½ lb.

SCALE OF POINTS

TYPE	25
HEAD	20
COLOUR	15
LEGS AND FEET	10
SIZE	10
QUALITY	10
CONDITION	10
	100

Serious Defects: Comb with side spikes. Black, white or yellow shanks or toes. White in face. White in plumage or undercolour of the Black. Straw-colour or cream tinge in the White. Any deformity.

CAMPINE

ORIGIN: Belgium
CLASSIFICATION: Light
EGG COLOUR: White

THE Campine is of ancient Belgian lineage, famous for producing the finest winter milk chickens. There the breed was called the Braekel. Although not so widely kept for egg production as formerly, it has achieved some notoriety in that it has played a major role in the work of auto-sex-linkage. The curiosity of the genetic workers at Cambridge was aroused by the fact that the Campine is not, as might be supposed, a barred breed like the Barred Rock. This difference led to the making of the autosexing Cambar.

GENERAL CHARACTERISTICS: MALE

Carriage: Alert and graceful.

Type: Body broad, close and compact. Back rather long, narrowing to the tail. Breast full and round. Wings large and neatly tucked. Tail long, carried well out and with broad and plentiful sickles and secondaries.

Head: Moderately long, deep, and inclined to width. Beak rather short. Eyes prominent. Comb single, or rose. The single upright, of medium size, evenly serrated, the back carried well out and clear of neck; free from excrescences. The rose firmly and evenly set, medium size and height, square front, tapering towards the back, leader straight and level with the surface, the top full of fine work and free from hollows. Face smooth. Ear-lobes inclined to almond shape, medium size, free from wrinkles. Wattles long and fine.

Neck: Moderately long and well covered with hackle feathers.

Legs and Feet: Legs moderately long. Shanks and feet free from feathers. Toes, four, slender and well spread.

FEMALE

With the exception of the single comb which falls gracefully over one side of the face, the general characteristics are similar to those of the male, allowing for the natural sexual differences.

COLOUR

Plumage, Gold Variety, Male and Female: Head and neck hackle rich gold, not a washed-out yellow. Remainder beetle-green barring on rich gold ground. Every feather must be barred in a transverse direction with the end gold, the bars being clear and with well-defined edges, running across the feather so as to form, as near as possible, rings around the body and three times as wide as the ground (gold) colour. On the breast and underparts of body the barrings should be straight or slightly curved; on the back, shoulders, saddle and tail they may be of a V-shaped pattern, but

Campine
Silver Male and
Female

preferably straight. The male should be furnished with properly developed saddle hackles.

Plumage, Silver Variety, Male and Female: Head and neck hackle pure white. Remainder beetle-green barring on pure white ground, the markings being as in the Gold.

In both sexes: Beak horn. Eyes dark brown with black pupil. Comb, face and wattles bright red. Ear-lobes white. Legs and feet leaden-blue. Toe nails horn.

Standard Weights: Male 6 lb. Female 5 lb.

SCALE OF POINTS

SIZE	10
HEAD (COMB 5, EYES 5, LOBES 5)	15
COLOUR (NECK HACKLE 12, SHEEN 10)	22
TAIL (DEVELOPMENT AND CARRIAGE)	8
LEGS AND FEET	5
MARKINGS	30
CONDITION	10
	100

Interpretation: The ideal is a bird clearly, distinctly and evenly barred all over with the sole exception of its neck hackle, which should be of the ground colour of the body. So that, taking the five main points of a bird—viz. Neck hackle, top (including back, shoulders and saddle), tail, wing and breast—each is of as much importance as another; and judges are requested to bear in mind that a specimen excelling in one or two particulars but defective in others should stand no chance against one of fair average merit throughout. Special attention should be paid to size, type, and fullness of front in breeding and judging Campines.

Serious Defects: Bars and ground colour of equal width. Ground colour pencilled. Side spikes (or sprigs) on comb. Legs other than leaden-blue. White in face. Red eyes. Feather or fluff on shanks. Any deformity.

COCHIN

ORIGIN: Asiatic
CLASSIFICATION: Heavy
EGG COLOUR: Tinted

THE Cochin, as we know it today, originally came from China in the early " 50's ", where it was known as the Shanghai, and later still as the Cochin-China. The breed created a sensation in this country in poultry circles because of its immense size, and table properties. Moreover, it was an excellent layer. It was developed, however, for wealth of feather and fluff for exhibition purposes to the extent that its utility characteristics were neglected, if not made impossible, in winning types.

GENERAL CHARACTERISTICS: MALE

Carriage: Rather forward, high at stern, and dignified.

Type: Body large and deep. Back broad and very short. Saddle very broad and large with a gradual and decided rise towards the tail forming an harmonious line with it. Breast broad and full, as low down as possible. Wings small and closely clipped up, the flights being neatly and entirely tucked under the secondaries. Tail small, soft, with as little hard quill as possible and carried low or nearly flat.

Head: Small. Beak rather short, curved and very stout at base. Eyes large and fairly prominent. Comb single, upright, small perfectly straight, of fine texture, neatly arched and evenly serrated, free from excrescences. Face smooth, as free as possible from feathers or hairs. Ear-lobes sufficiently developed to hang nearly or quite as low as the wattles, which are long, thin and pendent.

Neck: Rather short, carried somewhat forward, handsomely curved, thickly furnished with hackle feathers which flow gracefully over shoulders.

Legs and Feet: Thighs large and thickly covered with fluffy feathers standing out in globular form; hocks entirely covered with soft curling feathers, but as free as possible from any stiff quills (vulture hocks). Shanks short and thick, wide apart and heavily feathered down the outside, the feathering to start from the hock and continue to the ends of the middle and outer toes.   Toes, four, large, straight and well spread.

General Shape: Massive and deep.

FEMALE

With certain exceptions the general characteristics are similar to those of the male, allowing for the natural sexual differences. Comb and wattles as small as possible. The body more square than the male's and the shoulders more prominent. The back very flat, wide and short, with the cushion exceedingly broad, full and convex, rising from as far forward as possible and almost burying the tail. Wings nearly buried in abundant body feathering and the tail very small. Breast full, as low as possible. General shape is " lumpy ", massive and square. Carriage is forward, high at cushion, with a matronly appearance.

COLOUR

Plumage, Black Variety, Male and Female: Rich black, well glossed, free from golden or reddish feathers.

In both sexes: Beak yellow, horn or black. Comb, face, ear-lobes and wattles bright red. Eyes bright red, dark red, hazel or nearly black. Legs dusky yellow or lizard.

Plumage, Blue Variety, Male: Hackle, back and tail level shade of rich dark blue free from rust, sandiness or bronze. Remainder even shade of blue free from lacing on breast, thighs or fluff and free from rust, sandiness or bronze.

Plumage, Female: One even shade of blue free from lacing; pigeon blue preferred.

In both sexes: Beak yellow, horn or yellow slightly marked with horn. Comb, face, ear-lobes and wattles bright red. Eyes dark. Legs and feet blue, with yellow tinge in pads.

Plumage, Buff Variety, Male: Breast and underparts any shade of lemon-buff, silver-buff or cinnamon provided it is even and free from mottling. Head, hackle, back, shoulders, wings, tail and saddle may be any shade of deeper and richer colour which harmonises well—lemon, gold, orange or cinnamon—wings to be perfectly sound in colour and free from mealiness. White in tail very objectionable.

Plumage, Female: Body all over any even shade, free from mottled appearance. Hackle of a deeper colour to harmonise, free from black pencilling or cloudiness, cloudy hackles being especially objectionable. Tail free from black.

Cochin
White Male,
Blue Female

In both sexes: Beak rich yellow. Comb, face, ear-lobes and wattles brilliant red. Eyes to match plumage as nearly as possible, but red eyes preferred although rare. Legs bright yellow with shade of red between the scales.

Plumage, Cuckoo Variety, Male and Female: Dark blue-grey bars or pencilling (across the feather) on blue-grey ground, the male's hackle free from golden or red tinge, and his tail free from black or white feathers.

In both sexes: Beak rich bright yellow, but horn permissible. Comb, face, ear-lobes and wattles as in the Black. Eyes bright red. Legs brilliant yellow.

Plumage, Partridge and Grouse Varieties, Male: Neck and saddle hackle rich bright red or orange-red, each feather with a dense black stripe. Back, shoulder coverts, and wing bow rich red, of a more decided and darker shade than the neck. Wing coverts green-black, forming a wide and sharply cut bar across the wing; secondaries rich bay outside and black inside, the end of every feather black; primaries very dark bay outside and dark inside. Saddle rich red or orange-red, the same colour as, or one shade lighter than, the neck. Remainder glossy black, as intense as possible, white in tail objectionable.

Plumage, Female: Neck bright gold, rich gold, or orange-gold, with a broad black stripe in each feather, the marking extending well over the crown of the head Remainder (including leg feathering) brown distinctly pencilled in crescent form with rich dark brown or black, the pencilling being perfect and solid up to the throat.

In both sexes: Beak yellow or horn. Comb, face, ear-lobes and wattles as in the Black. Eyes bright red. Legs yellow, but may be of a dusky shade.

Plumage, White Variety, Male and Female: Pure white, free from any straw or red shade.

In both sexes: Beak rich bright yellow. Comb, face, ear-lobes and wattles as in the Black. Eyes pearl or bright red. Legs brilliant yellow.

Standard Weights: Cock 10 to 13 lb; cockerel 8 to 11 lb. Hen 9 to 11 lb; pullet 7 to 9 lb.

SCALE OF POINTS

FEATHERING (CUSHION 8, FLUFF 7, TAIL 5,
    HACKLE 5, LEGS 10)...................... 35

Continued on next page

Cochin Scale of Points—continued

COLOUR (OR MARKINGS IN CUCKOO OR
    PARTRIDGE)......................... 20
SIZE................................. 15
HEAD (EAR-LOBES 5)................... 15
TYPE................................. 10
CONDITION............................ 5
                                —
                              100

Serious Defects: Primary wing feathers twisted on their axes. Utter absence of leg feather. Badly twisted or falling comb. Legs other than yellow or dusky yellow, except in Blacks and Blues. Black spots in Buffs. Brown mottling (if conspicuous) in Partridge males, or pale breasts destitute of pencilling in Partridge females. White or black feathers in Cuckoos. Crooked back, wry tail, or any other deformity.

CROAD LANGSHAN

ORIGIN: Asiatic
CLASSIFICATION: Heavy
EGG COLOUR: Brown

THE first importation of Langshans into this country was made by Major Croad, and as with other Asiatic breeds, controversy centred around it. Already there was the Black Cochin and then the Black Langshan, some contending both were one breed, and others that they were quite separate Chinese breeds. As developed here the breed was called the Croad Langshan after the name of the importer. In 1904 a Croad Langshan club was formed to maintain the original stamp of bird. The Modern Langshan has been developed along different lines, and in consequence the two types are shown in separate classes at shows.

GENERAL CHARACTERISTICS: MALE

Carriage: Graceful, well balanced, active and intelligent.

Type: Back of medium length, broad and flat across shoulders, the saddle well filling the angle between the back and the tail as seen in profile. In the male the back should appear shorter than in

53

the female. Breast broad, deep and full (fuller in old bird) with long breast bone, the keel slightly rounded. Wings carried high. Tail fan-shaped, well spread to right and left and carried rather high; it should be level with the head when the bird stands in position of attention; side hangers plentiful, and two sickle feathers on each side projecting some six inches or more beyond rest. Abdomen capacious and resilient to touch, with fine pelvic bones. Saddle rather abundantly furnished with hackles.

Head: Carried well back, small for size of bird, full over the eyes. Beak fairly long and slightly curved. Eyes large and intelligent. Comb single, upright, straight, medium or rather small, free from side sprigs, thick and firm at the base, becoming rather thin, fine and smooth in texture, evenly serrated with five or six spikes (five preferred). Face free of feathers. Ear-lobes well developed, pendent, and fine in texture. Wattles fine in quality and rather small.

Neck: Of medium length, with full neck hackle.

Legs and Feet: Legs sufficiently long to give a graceful carriage to the body which should be well balanced, an adult bird neither high nor low on leg. Thighs rather short but long enough to let the hocks stand clear of fluff, well covered with soft feathers. Shanks medium length, well apart, feathered down outer sides (neither too scantily nor too heavily). Toes four, long, straight and slender, the outer toe feathered.

Plumage: Rather soft, neither loose nor tight.

Table Merits: Size for table purposes must be a great consideration, consistent with type. Bone medium or rather fine, in due proportion to size, but subordinate to amount of meat carried. Male has higher proportion of bone than female. Skin, thin and white; flesh white.

FEMALE

Hocks need not show in adult as she carries more fluff than the male. Cushion fairly full but not obtrusive. Tail may have two feathers slightly curved and projecting about one inch beyond rest. In other respects similar to the general characteristics of the male, allowing for the natural sexual differences.

COLOUR

Plumage, Male and Female: Surface dense black with beetle-green gloss free from purple or blue tinge. Undercolour dark grey, darker in the female. White in foot feather characteristic and not a defect.

Croad Langshan
Male and Female

In both sexes: Beak light to dark horn, preferably light at tip and streaked with grey. Eyes brown, the darker the better, but not black (ideal is colour of ripe hazel nut, a Vandyke brown). Comb, face, wattles and ear-lobes brilliant red. Shanks bluish-black (bluish in adult birds, scales and toes nearly black in young birds) showing pink between scales especially on back and inner side of shank. In male bird intense red should show through the skin along outer side at base of shank feathers. Toes, the web and bottom of foot pinkish white, the deeper the pink the better; black spots on soles a serious fault. Toe nails white; dark colour or black a serious fault.

White Langshans (Croad Type): The general characteristics are the same as for the original Croad above. Plumage is pure white, and the beak light horn. Eyes, comb, wattles and legs are as in the original Croad. Serious defects are black or coloured feathers, and black tips to the feathers.

Standard Weights: Male 9 lb (min.). Female 7 lb (min.).

SCALE OF POINTS

TYPE AND CONDITION (SHAPE 15, CONDITION 10)	25
BODY (GIRTH 15, FRAME AND BONE 10)	25
PLUMAGE (COLOUR 15, FURNISHINGS AND FOOTINGS 10)	25
HEAD, FEET AND ABDOMEN (HEAD AND FEET 15, ABDOMEN AND PELVIS 10)	25
	100

Note: It is left to discretion of judge to penalise any bad fault to the extent of 25 points.

Disqualifications: Yellow legs or feet. Yellow in face at base of beak or in edge of eyelids. Five toes. Other than single comb. Permanent white in ear-lobes. Grey (light slate colour) in webbing of flights. Black or partly black soles of feet as distinct from black spots.

Not Objectionable (in stock birds and not seriously against birds in show pen, judges using their discretion): Purple or blue barring in few feathers only where others are of good colour. Dark red in few feathers in neck hackle or on shoulders of male. White (not grey) in flights and secondaries. White tips on head of adult female. White tips or edging on breast in chicken feathers. Moderate amount of feathering on middle toe.

Highly Objectionable (to be firmly discouraged): Appreciable amount of purple or blue barring. Decided purple or blue tinge. Light eye (make some allowance for age). Yellow iris. Wry or squirrel tail. Marked scarcity or absence of leg and foot feather.

DORKING

ORIGIN: British
CLASSIFICATION: Heavy
EGG COLOUR: Tinted

Its purely British ancestry makes the Dorking one of the oldest of domesticated fowls in lineage. A Roman writer, who died in A.D. 47, described birds of Dorking type with five toes, and no doubt such birds were found in England by the Romans under Julius Caesar. By judicious crossings, and by careful selection, the Darking or Dorking breed was established.

GENERAL CHARACTERISTICS: MALE

Carriage: Quiet and stately, with breast well forward.

Type: Body massive, long and deep, rectangular in shape when viewed sideways, and tightly feathered. Back broad and moderately long with full saddle inclined downward to the tail. Breast deep and well rounded with a long straight keel bone. Wings large and well tucked up. Tail full and sweeping carried well out (a " squirrel " tail being objectionable) with abundant side hangers and broad well-curved sickles.

Head: Large and broad. Beak stout, well proportioned and slightly curved. Eyes full. Comb single or rose. Either kind is allowed in Darks, single only in Reds and Silver-Greys, and rose only in Cuckoos and Whites. The single comb is upright, moderately large, broad at base, evenly serrated, free from thumb marks or side spikes. The rose is moderately broad and square-fronted, narrowing behind to a distinct and slightly upturned leader, the top covered with small coral-like points of even height, free from hollows. Face smooth. Ear-lobes moderately developed and hanging about one-third the depth of the wattles, which are large and long.

Neck: Rather short, covered with abundant hackle feathers falling well over the back, making it appear extremely broad at the base, and tapering rapidly at the head.

Legs and Feet: Legs short and strong. Thighs large and well developed but almost hidden by the body feathering. Shanks short, moderately stout and round (square or sinewy bone being very objectionable), free from feathers, the spurs set on the inner side and pointing inwards. Toes, five, large, round and hard ("spongy" feet to be guarded against), the front toes (three) long, straight and well spread, the hind toe double and the extra toe well formed, viz., the normal toe as nearly as possible in the natural position, and the extra one placed above, starting from close to the other, but perfectly distinct and pointing upwards.

FEMALE

The general characteristics are similar to those of the male, allowing for the natural sexual differences, except that the tail is carried rather closely. The single comb, too, falls over one side of the face.

COLOUR

Plumage, Cuckoo Variety, Male and Female: Dark grey or blue bands (barring) on light blue-grey ground, the markings uniform, the colours shading into each other so that no distinct line or separation of the colours is perceptible.

Plumage, Dark Variety, Male: Hackles (neck and saddle) white or straw more or less striped with black. Back various shades of white, black and white or grey, mixed with maroon or red (bronze objectionable). Wing bows white, or white mixed with black or grey; coverts (or bar) black glossed with green; secondaries outer web white, inner black. Breast and underparts jet black; white mottling not permissible. Tail, richly glossed black, and a little white on primary sickles is permissible, but white hangers decidedly objectionable.

Plumage, Female: Neck hackle white or pale straw, striped with black or grey-black. Breast salmon-red, each feather tipped with dark grey verging on black. Tail nearly black, the outer feathers slightly pencilled. Remainder of plumage nearly black, or approaching a rich dark brown, the shaft showing a cream-white, each feather slightly pale on the edges, except on the wings, where the centre of the feather is brown-grey covered with a small rich marking surrounded by a thick lacing of the black, and free from red. Another successful colour is every feather over the body pencilled a brown-grey in the centre, with lacing round, and the breast as described above.

58

Dorking
Silver-grey Male,
Dark Female

Plumage, Red Variety, Male: Hackles (neck and saddle) bright glossy red. Back and wing-bows dark red. Remainder of plumage jet black glossed with green.

Plumage, Female: Hackle bright gold heavily striped with black. Tail and primaries black or very dark brown. Remainder of plumage red-brown, the redder the better, each feather more or less tipped or spangled with black, and having a bright yellow or orange shaft.

Plumage, Silver-grey Variety, Male: Hackles (neck and saddle) silver-white free from straw tinge or marking of any kind. Back, shoulder coverts and wing bow silver-white free from striping. Wing coverts lustrous black with green or blue gloss; primaries black with a white edge on outer web; secondaries white on outer and black on inner web, with a black spot at the end of each feather, the corner of the wing when closed appearing as a bar of white with a black upper edge. Remainder of plumage deep black, free from white mottling or grizzling, although in old males a slight grizzling of the thighs is not objectionable.

Plumage, Female: Hackle silver-white, striped with black. Breast robin-red or salmon-red ranging to almost fawn, shading off to ash-grey on the thighs. Body clear silver-grey, finely pencilled with darker grey (the pencilling following the outer line of the feather), free from red or brown tinge or black dapplings.
Note: The effect may vary from soft dull grey to bright silver-grey, an old fashioned grey slate best describing the colour. Tail darker grey, inside feathers black.

Plumage, White Variety, Male and Female: Snow white, free from straw tinge.

In both sexes: Beak white or horn, dark horn permissible in the Dark. Eyes bright red. Comb, face, wattles and ear-lobes brilliant red. Legs and feet (including nails) a delicate white with a pink shade.

Standard Weights: Cock 10 to 14 lb; cockerel 8 to 11 lb. Hen 8 to 10 lb.

SCALE OF POINTS

	Dark	Silver-grey or Red	Cuckoo or White
SIZE	28	18	15
TYPE	20	12	20
COLOUR	12	24	15

Continued on next page

60

Dorking Scale of Points—continued

FIFTH TOE	10	10	15
CONDITION	12	12	10
HEAD	10	16	17
FEET, CONDITION OF	8	8	8
	100	100	100

Serious Defects: Total absence of fifth toe. Legs other than white or pink-white, or with any sign of feathers. Spurs outside the shank. Single comb in Cuckoo or White. Rose comb in Red or Silver-grey. White in breast or tail of Silver-grey male. Any coloured feathers in White. Very long legs. Crooked or much swollen toes. Bumble feet. Any deformity.

FAVEROLLE

ORIGIN: French
CLASSIFICATION: Heavy
EGG COLOUR: Tinted

ORIGINATED in the village of Faverolle, in Northern France, this breed was created for its dual-purpose qualities. Its make-up includes such breeds as the Dorking, Houdan and Cochin, while Light Brahma blood as well as that of the Malines may be seen in some of the varieties. Imported into Great Britain in 1886, producers of table chickens crossed it freely with the Sussex, Orpington and Indian Game.

GENERAL CHARACTERISTICS: MALE

Carriage: Active and alert.

Type: Body deep, thick and " cloddy ". Back fairly long, flat and " square ", i.e., very broad across shoulders and saddle. Breast broad, keel-bone very deep and well forward in front, but not too rounded (a hollow breast is very objectionable). Wings small, prominent in front, carried closely. Tail moderately long, somewhat upright, and with broad feathers (flowing tail, either low or on a level with the back, is very objectionable).

Head: Broad, flat and short, free from crest. Beak short, stout. Eyes prominent. Comb single, medium size, upright, with four to six serrations, smooth and free from coarseness or any side work. Face muffled; muffling full, wide, short and solid. Ear-lobes and wattles small, of fine texture, and partly concealed by the muffles.

Neck: Short and thick, especially near the body, which it should be well let into.

Legs and Feet: Legs short and stout. Thighs wide apart. Shanks straight and medium length with width between them, and sparsely feathered to the outer toe. Narrowness or any tendency to in-kneed is very objectionable. Toes, five, the front three long, straight and well spread, the outer toe sparsely feathered, the fourth toe (quite divided from the fifth) on the ground and well back, the fifth turned up the leg.

FEMALE

The general characteristics are similar to those of the male, allowing for the natural sexual differences, with the following exceptions. Comb much smaller in proportion, back longer in proportion, neck straighter, keel bone longer and deeper, and the tail carried midway between upright and drooping.

COLOUR

Plumage, Black Variety, Male and Female: Black showing rich beetle-green sheen, free from purple bars.

In both sexes: Beak black. Eyes black or brown. Comb red. Face, wattles and ear-lobes red, partly concealed by muffling. Legs and feet black.

Plumage, Blue Variety, Male: Head, muffling, hackles, back, tail and wing bows a uniform dark blue. Remainder rich blue, each feather laced a dark shade.

Plumage, Female: Rich uniform blue, each feather laced a dark shade.

In both sexes: Beak, etc., as in the Black except that the legs and feet may also be blue.

Plumage, Buff Variety, Male and Female: Rich lemon-buff throughout.

In both sexes: Beak horn or white. Eyes grey or hazel. Comb red. Face, wattles and ear-lobes red, partly concealed by muffling. Legs and feet white.

Plumage, Ermine Variety, Male and Female: Head and neck plumage white striped with black, the centre of each feather entirely

Faverolle
Salmon Male,
White Female

surrounded by a white margin.   Wings white with black in flights. Tail black.   Remainder pure white.

In both sexes: Beak, etc., as in the Buff.

Plumage, Salmon Variety, Male: Beard and muff black.   Hackles straw.   Back, shoulders and wing bows bright cherry mahogany. Breast, thighs, underfluff, tail and shank feathering black.   Wing bar black; primaries black; secondaries white outer edge, black inner edge and at tips.

Plumage, Female: Beard and muff creamy-white.   Breast, thighs and fluff cream.   Remainder wheaten brown; head and neck striped with dark shade of the same colour (free from black) and wings softer and lighter than back.   Primaries, secondaries and tail wheaten brown.

In both sexes: Beak, etc., as in the Buff.

Plumage, White Variety, Male and Female: Pure white.

In both sexes: Beak, etc., as in the Buff.

Standard Weights: Cock 8 to 10 lb; cockerel 6½ to 9 lb.   Hen 6½ to 8½ lb; pullet 6 to 8 lb.

SCALE OF POINTS

UTILITY QUALITIES, SIZE AND CONDITION	25
TYPE	25
COLOUR	20
BEARD AND MUFFLING	15
FORMATION OF FEET AND TOES	5
FOOT FEATHER	5
COMB	5
	100

Serious Defects: Skin and legs other than white (except in the Black and the Blue).   Absence of muffling.   Featherless shanks and outer toes.   Other than five toes on each foot.   Hollow breast.   White or brassiness in hackle, wing or saddle, or purple barring, or white in foot-feather in the Blue.   Mealiness of general colour, or white in tail, wings and undercolour of the Buff.   Other than white legs, smuttiness on back, in the Ermine.   Brassiness on wings of White male.   Any bodily deformity.

FRIZZLE

Origin: Asiatic

Classification: Heavy

Egg Colour: White or tinted

THE Frizzle, a purely exhibition breed, is of Asiatic origin, and is notable for its quaint feather formation, each feather curling towards the head of the bird. More popular in bantams than in large fowls.

GENERAL CHARACTERISTICS: MALE

Carriage: Strutting and erect.

Type: Body broad and short. Breast full and rounded. Wings long. Tail rather large, erect, full but loose, with full sickles and plenty of side hangers.

Head: Fine. Beak short and strong. Eyes full and bright. Comb single, medium sized and upright. Face smooth. Ear-lobes and wattles moderate size.

Neck: Of medium length, abundantly frizzled.

Legs and Feet: Legs of medium length. Shanks free from feathers. Toes, four, rather thin, and well spread.

Plumage: Moderately long, broad and crisp, each feather curled towards the bird's head, and the frizzling as close and abundant as possible.

FEMALE

The general characteristics are similar to those of the male, allowing for the natural sexual differences, except that the comb is much smaller and the neck is not so abundantly frizzled.

COLOUR

Plumage, Male and Female: Black, blue, buff or white, a pure even shade throughout in the " self-coloured " varieties; Columbian as in Wyandotte; Duckwing, Black-red, Brown-red, Cuckoo, Pile, and Spangle as in Old English Game; Red as in Rhode Island Red.

Frizzle
Grey Male,
Red Female

In both sexes: Beak yellow in the Buff, Columbian, Pile, Red and White varieties; white in the Spangle, Black-red and Cuckoo; and dark willow, black or blue in other varieties. Eyes red. Comb, face, wattles and ear-lobes bright red. Legs and feet to correspond with the beak.

Standard Weights: Cock 8 lb; cockerel 7 lb. Hen 6 lb; pullet 5 lb.

SCALE OF POINTS

TYPE	25
COLOUR	25
CURL OF FEATHER	30
CONDITION	10
WEIGHT	10
	100

Serious Defects: Narrow feather. Want of curl. Long tail. Drooping comb. Other than single comb. White lobes. Deformity of any kind.

HAMBURGH

ORIGIN: North Europe
CLASSIFICATION: Light
EGG COLOUR: White

THE origin of the Hamburgh is wrapped in mystery. The Spangled were bred in Yorkshire and Lancashire three hundred years ago as Pheasants and Mooneys, and there is a book reference to Black Pheasants in the North of England in 1702. In its heyday the Hamburgh was a grand layer and must have played its part in the making of other laying breeds. However, its breeders directed it down purely exhibition roads, until today it is in few hands.

GENERAL CHARACTERISTICS: MALE

Carriage: Alert, bold and graceful.
Type: Body moderately long, compact, fairly wide and flat at the shoulders. Breast well rounded. Wings large and neatly tucked.

Tail long and sweeping, carried well up (but avoiding " squirrel " carriage), the sickles broad and the secondaries plentiful.

Head: Fine. Beak short, well curved. Eyes bold and full. Comb rose, medium size, firmly set, square-fronted, gradually tapering to a long, finely ended spike (or leader) in a straight line with the surface and without any downward tendency, the top level (free from hollows) and covered with small and smooth coral-like points of even height. Face smooth and free from stubby hairs. Ear-lobes smooth, round and flat (not concave or hollow), varying in size according to the variety. Wattles smooth, round and of fine texture.

Neck: Of medium length, covered with full and long feathers, which hang well over the shoulders.

Legs and Feet: Legs of medium length. Thighs slender. Shanks fine and round, free of feathers. Toes, four, slender and well spread.

FEMALE

The general characteristics are similar to those of the male, allowing for the natural sexual differences.

COLOUR

Plumage, Black Variety, Male and Female: Rich black, with a distinct green sheen from head to tail, and especially on sickle feathers and tail coverts. Any approach to bronze or purple tinge or barring to be avoided.

In both sexes: Beak black or dark horn. Eyes, comb, face and wattles red. Ear-lobes white. Legs and feet black.

Plumage, Gold Pencilled Variety, Male: Bright red bay or bright golden chestnut, except the tail, which is black, the sickle feathers and coverts being laced all round with a narrow strip of gold.

Plumage, Female: Ground colour similar to the general colour of the male, and, except on the hackle (which should be clear of all marking, if possible), each feather distinctly and evenly pencilled straight across with fine parallel lines of a rich green-black, the pencilling and the intervening colour to be the same width, while the finer and the more numerous on each feather the better.

In both sexes: Beak dark horn. Eyes, comb, face and wattles red. Ear-lobes white. Legs and feet lead-blue.

Plumage, Silver Pencilled Variety, Male and Female: Except that the

Hamburgh
Silver Spangled Male
and Female

ground colour, and in the male the tail lacings, are silver, this variety is similar to the Gold Pencilled.

Plumage, Gold Spangled Variety, Male: Ground colour rich bright bay or mahogany; striping, spangling, tipping and tail rich green-black. Hackles and back, each feather striped down the centre. Wings, bow dagger-shaped tips at the end of each feather; bars (two), rows of large spangles, running parallel across each wing with a gentle curve, each bar distinct and separate; secondaries tipped with large round spangles, forming the " steppings ". Breast and underparts, each feather tipped with a round spot or spangle, small near the throat, increasing in size towards the thighs, but never so large as to overlap.

Plumage, Female: Ground colour and spangling are similar to those of the male. Hackle, wing bars and " steppings " as in the male. Tail coverts black, with a sharp lacing or edging of gold on each feather. Remainder, each feather tipped with a spangle, as round as possible, and never so large as to overlap, the spangling commencing high up the throat.

In both sexes: Beak, eyes, comb, face, wattles, ear-lobes, legs and feet as in the pencilled varieties.

Plumage, Silver Spangled Variety, Male: Ground colour pure silver; spangling and tipping rich green-black. Hackles, shoulders and back, each feather marked with small, dagger-like tips. Wing, bow dagger-shaped tips, increasing in size until they merge into what is known as the third bar; bars (two) and secondaries, breast and underparts similarly marked to those of the Gold Spangled variety. Tail ending with bold half-moon-shaped spangles; sickles with large round spangles at the end of each feather; coverts similar, though spangles not so big.

Plumage, Female: Ground colour and spangling similar to those of the male. Hackle marked from the head with dagger-shaped tips, which gradually increase in width until they merge into the spangles at the bottom. Wings, secondaries as in the male, bars similar to those of the Gold Spangled female. Tail, each feather with a half-moon-shaped spangle at the end. Coverts reaching half-way up the true tail feathers to form a row across the tail (each side) of round spangles. Remainder marked as in the Gold female.

In both sexes: Beak, eyes, comb, face, wattles, ear-lobes, legs and feet as in the pencilled varieties.

Standard Weights: Male about 5 lb. Female about 4 lb.

Hamburgh
Black Male, Gold
Pencilled Female

SCALE OF POINTS

Black Variety

	Male	Female
TYPE, STYLE AND CONDITION........	15	15
HEAD (COMB, FACE AND EAR-LOBES 15 EACH)...........................	45	45
COLOUR (LEGS 5).....................	25	35
TAIL.................................	15	5
	100	100

Pencilled Variety

	Male	Female
TYPE, STYLE, CONDITION............	10	10
HEAD (COMB, EAR-LOBES AND FACE)..	25	20
COLOUR (INCLUDING LEGS)..........	30	10
MARKINGS (BACK AND CUSHION 15, BREAST AND THIGHS 15, TAIL 15, WINGS 10, NECK HACKLE 5).......	—	60
TAIL MARKINGS......................	35	—
	100	100

Gold Spangled Variety

	Male and Female
TYPE, STYLE AND CONDITION........	10
HEAD (COMB 10, EAR-LOBES 5, FACE 5)	20
COLOUR (INCLUDING LEGS)..........	10
MARKINGS (BACK AND SADDLE 15, BREAST AND THIGHS 15, WINGS 15, NECK HACKLE 10, TAIL 5).........	60
	100

Silver Spangled Variety

	Male	Female
TYPE, STYLE AND CONDITION........	10	10
HEAD (COMB 10, EAR-LOBES 10, FACE 5)	25	—
HEAD (COMB 10, EAR-LOBES 5, FACE 5)	—	20
COLOUR (INCLUDING LEGS)..........	10	10
MARKINGS (TAIL 15, NECK-HACKLE 10, BACK AND SADDLE 10, BREAST AND THIGHS 10, WINGS 10).......	55	—
MARKINGS (BACK AND CUSHION 15, NECK HACKLE 15, TAIL 10, BREAST AND THIGHS 10, WINGS 10).......	—	60
	100	100

Serious Defects: White face. Single comb. Red ear-lobes. Squirrel or wry tail. Any other deformity.

HOUDAN

ORIGIN: French
CLASSIFICATION: Light
EGG COLOUR: White

INTRODUCED into England in 1850, the Houdan is one of the oldest French breeds, taking its name from the town of Houdan, and being developed for table qualities. Developed here it was once classified as a heavy breed, but today is included in the category of light, non-sitting breeds. It is one of the few breeds carrying a fifth toe, a semi-dominant feature when crossed with other breeds.

GENERAL CHARACTERISTICS: MALE

Carriage: Bold and active.

Type: Body broad, deep and lengthy, as in the Dorking. Tail full with the sickles long and well arched.

Head: Fairly large, with a decidedly pronounced protuberance on top, and crested. Crest full and compact, round on top and not divided or " split ", composed of feathers similar to those of the hackle, inclining slightly backwards fully to expose the comb, in no way obstructing the sight except from behind. Beak rather short and stout, well curved, and with wide nostrils. Eyes bold. Comb leaf type, somewhat resembling a butterfly placed at the base of the beak, fairly small, well defined, and each side level. Face muffled; muffling large, full, compact, fitting around to the back of the eyes and almost hiding the face. Ear-lobes small, entirely concealed by muffling. Wattles small and well rounded, almost concealed by beard.

Neck: Of medium length, with abundant hackle coming well down on the back.

Legs and Feet: Legs short and stout, well apart, free of feathers. Toes, five, similar to those of the Dorking.

FEMALE

The general characteristics are similar to those of the male, allowing for the natural sexual differences, with the exception of the crest, which is full, compact and globular, not in any way obstructing the sight except from behind, and with the comb visible.   Tail fairly full.

COLOUR

Plumage, Male and Female: Glossy green-black ground with pure white mottles, the mottling to be evenly distributed, except on the flights and secondaries, and in the male the sickles and tail coverts, which are irregularly edged with white.   *Note:* In young Houdans black generally preponderates, but what mottling there is should be even and clear.   Mottling becomes gayer with age.

In both sexes: Beak horn.   Eyes red.   Comb, face and wattles bright red.   Ear-lobes white or tinged with pink.   Legs and feet white mottled with lead-blue or black.

Standard Weights: Male 7 to 8 lb.   Female 6 to 7 lb.

Houdan Female

SCALE OF POINTS

	Male	Female
TYPE	12	10
SIZE	18	20
COMB	15	8
LEGS AND FEET	10	10
COLOUR	15	15
CREST	12	15
MUFFLING	8	12
CONDITION	10	10
	100	100

Serious Defects: Red or straw-coloured feathers. Loose crest obstructing the sight. Spur outside the shank. Feathers on shanks or toes. Other than five toes on each foot. Any deformity.

INDIAN AND JUBILEE INDIAN GAME

ORIGIN: British
CLASSIFICATION: Heavy
EGG COLOUR: Tinted

To Cornwall must go the credit for giving us the Indian Game. Breeds used in the make-up were the Red Aseel, Old English Black-breasted Red Game, and the Malay. The breed has been developed for its abundant quantity of breast meat, in which respect no other breed can equal it. When large table birds were the most popular in this country Indian Game males were chosen as mates for hens of such table breeds as the Sussex, Dorking and Orpington, to produce extra large crosses. The hens chosen for mating belonged to breeds possessing white flesh and shanks. Jubilee Indian Game are similar to Indians, but the lacing is white; in Indians it is black. The two varieties are often interbred.

GENERAL CHARACTERISTICS: MALE

Carriage: Upright, commanding and courageous, the back sloping downwards towards the tail. A powerful and broad bird very active, sprightly and vigorous.

Type: Body very thick and compact and very broad at the shoulders, the shoulder butts showing prominently, but the bird must not be hollow-backed, the body tapering towards the tail. Back flat and broad at the shoulders, but the bird must not be flat sided. Elegance is required with substance. Breast wide, fairly deep and prominent, but well rounded. Wings short and carried closely to the body, well rounded at the points, closely tucked at ends and carried rather high in front. Tail medium length with short narrow secondary sickles and tail coverts, close and hard; carriage drooping.

Head: Rather long and thick, not so keen as in English Game, nor as thick as in the Malay; somewhat beetle browed but not nearly as much as in the Malay. Skull broad. Beak well curved and stout where set on the head giving the bird a powerful appearance. Eyes full and bold. Comb (in undubbed birds) pea type, i.e., three longitudinal ridges, the centre one being double height of those at sides, small closely set on the head. Ear-lobes and wattles smooth and of fine texture.

Neck: Of medium length and slightly arched; hackle short, barely covering base of the neck.

Legs and Feet: Legs very strong and thick. Thighs round and stout, but not as long as in the Malay. Shanks short and well scaled. The length of shank must be sufficient to give the bird a " gamey " appearance. Feet strong and well spread. Toes long, strong, straight, the back toe low and nearly flat on ground; nails well shaped.

Plumage: Short, hard and close.

Handling: Flesh firm.

FEMALE

The general characteristics are similar to those of the male, allowing for the natural sexual differences. Tail, however, which is well venetianed but close, is carried low but somewhat higher than the male's.

COLOUR

Plumage, Male: Head, neck, breast, underfluff, thighs, and tail black, with rich green glossy sheen or lustre, the base of the neck and tail hackles a little broken with bay or chestnut, which should be almost hidden by the body of the feathers. Shoulders and

76

Indian Game
Male and Female

wing bows green glossy black or beetle-green, slightly broken with bay or chestnut in the centre of the feather or shaft. Tail coverts green glossy black or beetle-green slightly broken with bay or chestnut in the base of the shaft. Back feathers green glossy black or beetle-green, also touched on the fine fronds at the end of the feathers with bay or chestnut which gives the sheen so much desired. When the wing is closed there is a triangular patch of bay or chestnut formed of the secondaries, which are green glossy black or beetle-green on the inner, and bay or chestnut on the outer web, and which when closed show only the bay in a solid triangle. The primaries, ten in number, are curved and of a deep black, except for about 2½ inches of a narrow lacing of light chestnut on the outer web.

Plumage, Female: The ground colour is chestnut brown, nut brown, or mahogany brown. Head, hackle and throat green glossy black or beetle-green. The pointed hackle that lies under the neck feathers green glossy black, or beetle-green with a bay or chestnut centre mark; the breast commencing on the lower part of the throat, expanding into double lacing on the swell of the breast, of a rich bay or chestnut, the inner or double lacing being most distinct, the belly and thighs being marked somewhat similarly and running off into a mixture of indistinct markings under the vent and swell of the thighs. The feathers of shoulders and back are somewhat smaller, enlarging towards the tail coverts and similarly marked with double lacing; the markings on wing bows and shoulders running down to the waist are most distinct of all, with the same kind of double lacing. Often in the best specimens there is an additional mark enclosing the base of the shaft of the feather and running to a point in the second or inner lacing. Tail coverts are seldom as distinctly marked, but have the same style of marking. Primary or flight feathers are black, except on inner frond or web which are a little coloured or peppered with a light chestnut. Secondaries are black on the inner web, while the outer web is in keeping with the general ground colour and is edged with a delicate lacing of green glossy black or beetle-green. Wing coverts which form the bar are laced like those of the body and often a little peppered. The black lacing should be metallic green, glossy black or beetle-green. This should appear embossed or raised.

In both sexes: Beak horn, yellow, or horn striped with yellow. Eyes from pearl to pale red. Face, comb, wattles and ear-lobes rich red. Legs rich orange or yellow, the deeper the better.

Standard Weights: Male 8 lb (min.). Female 6 lb (min.).

Jubilee Indian Game
Male and Female

SCALE OF POINTS

TYPE AND COLOUR (BODY AND THIGHS 10,
BACK, BREAST, WINGS, TAIL, LEGS, 8 EACH,
NECK 3)...................................... 53
CARRIAGE................................. 12
SIZE...................................... 10
HEAD (SKULL, EYES AND BROWS, 3 EACH;
BEAK, WATTLES, LOBES, COMB, 2 EACH) 17
CONDITION................................. 8
                                         ———
                                         100

Note : The Indian Game fowl is in no way allied to the English Game fowl. Hence it is not recognised as a true Game bird in the Fancy; that is, unless classes are specially provided for the breed it must compete in the " Any Other Variety " classes and not in those set aside for " Game," *vide* Poultry Club Show Rules.

Defects: Crooked breasts or toes. Flat shins. Rusty hackles. Bad shape. Heavy feathering. White in hackles. Smallness of size. Long legs and thighs. Twisted hackle.

Disqualifications: Male: Crooked back, beak and legs. Wry or squirrel tail, in-knees, bent legs and flat sides. Single or Malay comb. Red hackles. Additionally in the female too light, too dark or mealy ground colour, and defective markings.

JUBILEE INDIAN GAME

GENERAL CHARACTERISTICS: As for Indian Game.

COLOUR

Plumage, Male: Head, neck, breast, body, underfluff, thighs and tail white. Hackle feathers to have chestnut shaftings. Clear breasts are desirable. Wing bows and shoulders white, slightly broken with bay or chestnut. Wing primaries and secondaries white with bay markings. Triangular patch of bay or chestnut to show when wing is closed. Tail coverts white. Back, white touched with bay or chestnut.

Plumage, Female: Ground colour chestnut brown or mahogany. Head hackle and throat white. Breast, commencing on the lower part of the throat and expanding to double lacing on the swell of the breast, mahogany laced with white. The inner, or double lacing, to be most distinct. The underparts and thighs are marked somewhat similarly and run into a mixture of indistinct markings beneath the vent and swell of the thighs. Feathers of the shoulders and back somewhat small, enlarging towards the tail coverts similarly marked with the double lacing; often in the best specimens there is an additional mark enclosing the base of the shaft of the feather and running to a point in the second or inner lacing. The tail coverts are seldom as distinctly marked, but with the same style of marking. Wing primaries, white marked on inner web with chestnut. Secondaries, white inner web, chestnut outer web, edged with white. Main tail white. Remainder chestnut ground colour throughout, double laced with white, inner lacing should be quite distinct. Underparts and thighs may be less distinctly marked and wing coverts may be peppered.

In all other respects the Indian Game Standard should be followed.

In both sexes: Beak, eyes, comb and legs as described for Indians.

IXWORTH

ORIGIN: British
CLASSIFICATION: Heavy
EGG COLOUR: Tinted

ORIGINATED at the village of Ixworth in Suffolk, this breed was introduced mainly for its table qualities allied to steady egg production. It is clear from the general outline of the breed that the Indian Game played a part in its make-up.

GENERAL CHARACTERISTICS: MALE

Carriage: Alert, active and well balanced.

Type: Body deep, well rounded, fairly long but compact. Back long, flat, reasonably broad, without too prominent a slope to the tail. Breast broad, full, deep, well rounded, long and wide, low breast bone carried well forward; with unpronounced keel or keel point; well fleshed and rounded off for entire length. Wings strong, carried close, showing shoulder butts. Tail compact, of medium length and carried fairly low, the sickles close fitting.

Head: Broad and of medium length. Beak short and stout. Eyes full, prominent, keen expression, without heavy brows. Comb pea type. Face smooth and of fine texture. Ear-lobes and wattles medium size and fine texture.

Neck: Somewhat erect and of reasonable length. Hackle feathers short, close-fitting and in no way excessive or loose.

Legs and Feet: Legs well apart, and of reasonable length to ensure activity. Thighs well fleshed and of medium length. Shanks covered with tight scales, free from feathers. Toes, four, straight, well spread and firm stance. Bone characteristic of a first-class table bird.

Plumage: Short, silky and close fitting; fluff likewise.

FEMALE

The general characteristics are similar to those of the male, allowing for the natural sexual differences.

COLOUR

Plumage, Male and Female: White.

In both sexes: Beak white. Eyes red or bright orange. Comb,

Ixworth Male

face, wattles, and ear-lobes brilliant red.   Legs, feet, skin and flesh white.

Standard Weights: Cock 9 lb; cockerel 8 lb.   Hen 7 lb;   pullet 6 lb.

SCALE OF POINTS

TABLE MERITS............................ 40
SHAPE AND SIZE......................... 20
COLOUR (GENERAL)..................... 20
HEAD................................... 10
PLUMAGE AND CONDITION.............. 10
————
100
====

Serious Defects: Coarseness. Lack of activity. Loose feathers. Any point against table values or general usefulness. Any deformity.

JERSEY GIANT————————————————

ORIGIN: American
CLASSIFICATION: Heavy
EGG COLOUR: Tinted to Brown

ORIGINATED in New Jersey some seventy years ago, this American breed took the name of " Giant " because of the extra heavy weights that specimens could record.   Its make-up accounts for such poundage as it includes Black Java, Dark Brahma, Black Langshan and Indian Game.   When introduced into this country it was claimed for the breed that the birds were heavier than those of any other breed, that it was adaptable for farm range, and also for providing capons.   Earlier specimens were of exceptional weights.

GENERAL CHARACTERISTICS: MALE

Carriage: Bold, alert and well balanced.

Type: Body long, wide, deep and compact; smooth at sides, with long keel, smooth and moderately full fluff. Back rather long, broad, nearly horizontal, with a short sweep to the tail. Breast broad, deep and full, carried well forward. Wings medium sized, well folded, carried at the same angle as the body, the primaries and secondaries broad and overlapping in natural order when the wings are folded. Tail rather large, full, well spread, carried at an angle of 45 degrees above the horizontal, the sickles just sufficiently long to cover the main tail feathers, the coverts moderately abundant and of medium length, the main tail feathers broad and overlapping.

Head: Rather large and broad. Beak short, stout and well curved. Eyes large, round, full and prominent. Comb single, straight, upright, rather large and of fine texture, having six well defined and evenly serrated points, the blade following the shape of the neck. Face smooth and fine in texture. Ear-lobes smooth and rather large, extending down one-half of the length of the wattles. Wattles of medium size and fine texture, well rounded at the lower ends.

Neck: Moderately long, full and well arched.

Legs and Feet: Legs straight and set well apart. Thighs large, strong, of moderate length and well covered with feathers. Shanks strong, stout, medium length and free from feathers; scales fine; bone of good quality and proportionate to size of bird. Toes, four, of medium length, straight and well spread.

FEMALE

With the exception of the tail, which is well spread and carried at an angle of 30 degrees above the horizontal, the general characteristics are similar to those of the male, allowing for the natural sexual differences.

COLOUR

Plumage, Black Variety, Male and Female: The surface a lustrous green-black, and the undercolour slate or light grey.

In both sexes: Beak black, shading to yellow towards the tip. Eyes dark brown or hazel. Comb, face, wattles and ear-lobes red. Legs and feet black, with a tendency towards willow in adult birds, the underpart of the feet being yellow.

Plumage, White Variety, Male and Female: The surface and under-colour white.

In both sexes: Beak, willow (some yellow permissible at present). Eyes, dark brown to black. Comb, face, wattles and ear-lobes red. Legs and feet, willow, i.e., dark greenish-yellow, soles yellow. Skin nearly white.

Standard Weights: Cock 13 lb; cockerel 11 lb. Hen 10 lb; pullet 8 lb.

Jersey Giant
White Male,
Black Female

SCALE OF POINTS

SHAPE AND CARRIAGE	25
COLOUR	20
QUALITY	15
HEAD	10
SIZE AND SYMMETRY	10
CONDITION	10
LEGS AND FEET	10
	100

Serious Defects (in Whites): Smoky surface colour; side sprigs to comb; more than 2 lb below standard weight in mature stock. **Defects** (in Black and Whites): Overhanging eyebrows. Sluggishness. Coarseness. Excessive or superfine bone. (In Blacks): Black or dull black undercolour extending to the skin of the hackle, back, breast, or body and fluff. Positive white showing on surface of plumage. Other than yellow under the feet. More than 2 lb below the standard weight in mature stock.

LEGHORN

ORIGIN: Mediterranean
CLASSIFICATION: Light
EGG COLOUR: White

ITALY was the original home of the Leghorn, but the first specimens of the White variety reached this country from America about 1870, and of the Brown two years or so later. These early specimens weighed not more than 3½ lb each, but our breeders started to increase the body weight of the Whites by crossing in the Minorca and Malay, until birds were produced well up to the weights of the heavy breeds. In the post-war years, the utility and commercial breeders established a type of their own, and that is the one which is now favoured. In commercial circles the White

Leghorn
Black Male and
Female

Leghorn has figured prominently in the establishment of high egg-producing hybrids.

GENERAL CHARACTERISTICS: MALE

Carriage: Very sprightly and alert, but without any suggestion of stiltiness or in-kneed appearance. Well balanced.

Type: Body wide at the shoulders and narrowing slightly to root of tail. Back long and flat, sloping slightly to the tail. Breast round, full and prominent, carried well forward; breast bone straight. Wings large, tightly carried and well tucked up. Tail moderately full and carried at an angle of 45 degrees from the line of the back; full, sweeping sickles.

Head: Well balanced with fine skull. Beak short and stout, the point clear of the front of the comb. Eyes prominent. Comb single or rose. The single of fine texture, straight, and erect, moderately large but not overgrown, coarse or beefy, deeply and evenly serrated (the spikes broad at their base), extending well beyond the back of the head and following, without touching, the line of the head, free from " thumb marks " and side spikes, or twist at the back. The rose moderately large, firm (not overgrown so as to obstruct the sight), the leader extending straight out behind and not following the line of the head, the top covered with small coral-like points of even height and free from hollows. Face smooth, fine in texture and free from wrinkles or folds. Ear-lobes well developed and pendent, equally matched in size and shape, smooth, open and free from folds. Wattles long, thin and fine in texture.

Neck: Long, profusely covered with hackle feathers and carried upright.

Legs and Feet: Legs moderately long. Shanks fine and round—flat shins objectionable—and free of feathers. Ample width between legs. Toes, four, long, straight and well spread, the back toe straight out at rear. Scales small and close fitting.

Plumage: Of silky texture, free from woolliness or excessive feather.

Handling: Firm, with abundance of muscle.

FEMALE

With the exception of the single comb rising from a firm base and falling gracefully over either side of face without obstructing the

Leghorn
Buff Male,
Brown Female

sight, and the tail, which is carried closely and not at such a high angle, the general characteristics are similar to those of the male, allowing for the natural sexual differences.

COLOUR

Plumage, Black Variety, Male and Female: Rich green-black or blue-black, the former preferred and perfectly free of any other colour.

Plumage, Blue Variety, Male and Female: Even medium shade of blue from head to tail, free from lacing, a dark tint allowed in the hackles of the male, but no black, " sand " or any other colour than blue, and the more even the better.

Plumage, Brown Variety, Male: Head and hackle rich orange-red striped with black, crimson-red at the front of hackles below the wattles. Back, shoulder coverts and wing bow deep crimson-red or maroon. Wing coverts steel-blue with green reflections forming a broad bar across; primaries brown; secondaries deep bay on outer web (all that appears when wing is closed) and black on the inner web. Saddle rich orange-red with or without a few black stripes. Breast and underparts glossy black, quite free from brown splashes. Tail black glossed with green; any white in tail is very objectionable. Tail coverts black edged with brown.

Plumage, Female: Hackle rich golden-yellow, broadly striped with black. Breast salmon-red, running into maroon around the head and wattles, and ash-grey at the thighs. Body colour rich brown, very closely and evenly pencilled with black, the feathers free from light shafts, and the wings free from any red tinge. Tail black, outer feathers pencilled with brown.

Plumage, Buff Variety, Male and Female: Any shade of buff from lemon to dark, at the one extreme avoiding washiness and at the other a red tinge; the colour to be perfectly uniform, allowing for greater lustre on the hackle feathers and wing bow of the male.

Plumage, Cuckoo Variety, Male and Female: Light blue or grey ground, each feather barred across with bands of dark blue or grey, the markings to be uniform; the barring shading into the ground colour not cleanly cut but sharp enough to keep the two colours distinct.

Plumage, Golden Duckwing Variety, Male: Neck hackle rather light yellow or straw, a few shades deeper at the front below the wattles, the longer feathers striped with black. Back deep rich gold. Saddle and saddle hackle deep gold, shading in hackle to pale gold.

Leghorn
White and Exchequer
Males

Shoulder coverts bright gold or orange, solid colour (an admixture of lighter feathers is very objectionable). Wing bows the same as the shoulder coverts; coverts metallic blue (blue-violet) forming an even bar across the wing, sharp, cleanly cut and not too broad; primaries black, with white edging on the outer web; secondaries white outer web (all that appears when the wing is closed), black inner and end of feather. Breast black with green lustre. Tail black, richly glossed with green-grey fluff at the base.

Plumage, Female: Head grey (a brown cap is very objectionable). Hackle white, each feather sharply striped with black or dark grey (a light tinge of yellow in the ground colour admitted). Breast and undercolour bright salmon-red (this point is very important), darker on throat and shaded off to ash-grey or fawn on the underparts. Back, wings, sides and saddle dark slate-grey, finely pencilled with darker grey or black. Tail grey, slightly darker than the body colour, inside feathers dull black or dark grey.

Plumage, Silver Duckwing Variety, Male: Neck hackle silver-white, the long feathers striped with black. Back, saddle and saddle hackle silver-white. Shoulders and wing bow silver-white, as solid as possible (any admixture of red or rusty feathers very objectionable). Wing coverts metallic blue (blue-violet) forming an even bar across the wing, which should be sharp and clearly cut, and not too broad; primaries black with white edging on outer parts; secondaries white outer edge (all that appears when the wing is closed), black inner and end of feathers. Thighs and underparts black. Tail black richly glossed with green, grey fluff at the base.

Plumage, Female: Head silver-white. Hackle silver-white, each feather sharply striped with black or dark-grey. Breast and underparts light salmon or fawn, darker on throat and shaded off to ash-grey on underparts. Back, wings, sides and saddle clear delicate silver-grey or French grey, without any shade of red or brown, finely pencilled with dark grey or black (purity of colour very important). Tail grey, slightly darker than the body colour, with the inside feathers a dull black or dark grey.

Plumage, Exchequer Variety, Male and Female: Black and white evenly distributed with some white in the undercolour, the white of the surface colour in the form of a large blob as distinct from V-shaped ticking. Wings and tail to appear white and black evenly distributed.

Plumage, Mottled Variety, Male and Female: Black with white tips to each feather, the tips as evenly distributed as possible. Black to predominate and to have a rich green sheen.

Plumage, Partridge Variety, Male and Female: This colour is fully described under Wyandotte Bantams and need not be repeated here.

Plumage, Pile Variety, Male: Neck hackle bright orange. Back and saddle rich maroon. Shoulders and wing bows dark red. Secondaries dark chestnut outer web (all that appears when the wing is closed) and white inner. Remainder white.

Plumage, Female: Neck white tinged with gold. Breast deep salmon-red shading into white thighs. Remainder white.

Plumage, White Variety, Male and Female: Pure white free from straw tinge.

In both sexes: Beak yellow or horn. Eyes red. Comb, face and wattles bright red. Ear-lobes pure opaque white (resembling white kid) or cream, the former preferred. Legs and feet yellow or orange.

Standard Weights: Cock 7½ lb; cockerel 6 to 6½ lb. Hen 5½ lb; pullet 4½ to 5 lb.

SCALE OF POINTS

Black Variety

COMB	12
EAR-LOBES, FOLDED, WRINKLED OR STAINED WITH RED	15
COLOUR	25
LEGS	8
CONDITION	10
SIZE	15
SYMMETRY	15
	100

Serious Defects: Dark legs or eyes.

Blue Variety

COLOUR	30
TYPE	20
HEAD (COMB 10, LOBES 10)	20
LEGS	10
CONDITION	10
SIZE	10
	100

Serious Defects: Any lacing or barring on plumage. Dark eyes.

Brown Variety

COMB, TOO LARGE, TOO SMALL, BADLY SHAPED OR IMPERFECT	12
LOBE, FOLDED OR WRINKLED	6
LOBE, STAIN OF RED	10
COLOUR	20
HACKLE	10
SIZE	15
SYMMETRY	15
CONDITION	12
	100

Serious Defect: White feathers.

Buff Variety

TYPE	20
HEAD (COMB 12, LOBES 15)	27
COLOUR	20
SIZE	15
CONDITION	10
LEGS	8
	100

Cuckoo, Duckwing, and Pile Varieties

TYPE	25
COLOUR	20
HEAD (COMB 5, LOBES 5, BEAK AND EYES 5)	15
SIZE	15
TEXTURE OR QUALITY	10
CONDITION	10
LEGS	5
	100

Exchequer Variety

INDICATIONS OF PRODUCTIVENESS	30
TYPE AND SIZE	20
COLOUR	20
HEAD	10
LEGS	10
CONDITION	10
	100

Serious Defects: Entirely dark undercolour.   Black spots on legs.

94

White Variety

TYPE	25
HEAD	20
COLOUR	20
SIZE	15
CONDITION	10
LEGS	10
	100

Serious Defects, all varieties (for which a bird should be passed): *Single Comb*—Male's comb twisted or falling over, or female's erect. Ear-lobes red. Any white on face. Legs other than yellow or orange. Side sprigs on comb. Wry or squirrel tail, or any bodily deformity. *Rose Comb*—Comb other than rose or such as to obstruct sight. Ear-lobes red. White in face. Wry or squirrel tail or any bodily deformity. Legs other than orange or yellow.

MALAY

ORIGIN: Asiatic
CLASSIFICATION: Heavy
EGG COLOUR: Tinted

AT the first poultry show in England in 1845 the Malay had its classification, and in the first British Book of Standards of 1865 descriptions were included of both the Black-red and the White Malay. One of the oldest breeds, the Malay, reached this country as early as 1830, and our breeders developed it, particularly in Cornwall and Devon. Today it is not widely bred, and is a purely exhibition breed.

GENERAL CHARACTERISTICS: MALE

Carriage: Fierce, gaunt, very erect, high in front, drooping at stern, straight at the hock, and a hard, clean, cut-up appearance from behind.

Type: Body wide-fronted, short, and tapering; broad and square shoulders, the wing butts prominent, well up, and devoid of feathers at the point. Back short, sloping, and with convex outline, the saddle narrow and drooping. Breast deep and full, generally devoid of feathers at the point of the keel. Wings large, strong, carried high and closely to the sides. Tail of moderate length, drooping but not whipped, the sickles narrow and only slightly curved. The outline of the neck hackle, back, and tail (upper feathers) should form a succession of curves at nearly equal angles.

Head: Very broad with well-projecting or over-hanging (beetle) eyebrows, giving a cruel and morose expression. Beak short, strong, and curved downwards. The profile of the skull and the beak approaches in shape a section of a circle. Eyes deep-set. Comb shaped like a half-walnut, small, set well forward, and as free as possible from irregularities. Face smooth. Ear-lobes and wattles small.

Neck: Long and upright, with a slight curve, thick through from gullet to back of skull, the bare skin of the throat showing some way down the neck; the hackle full at the base of skull, but very short and scanty elsewhere.

Legs and Feet: Legs long and massive, set well on the front of the body. Thighs muscular with very little feather, leaving the hocks perfectly exposed. Shanks free of feathers, beautifully scaled, flat at the hocks and gradually rounding to the spurs, which should have a downward curve. Toes, four, long, and straight, with powerful nails, the fourth (or hind) toe close to the ground.

Handling: Firm fleshed and muscular.

Plumage: In all varieties is short and scanty, hard and narrow.

FEMALE

With the exception of the tail, which is carried slightly above the horizontal line, well " played " as if flexible at the joint, rather short and square and neither fanned nor whipped, the general characteristics are similar to those of the male, allowing for the natural sexual differences.

COLOUR

Plumage, Black Variety, Male and Female: Glossy black all over with brilliant green and purple lustre, the green predominant, free from brassy or white feather.

Plumage, Black-red Variety, Male: Hackle, saddle, back and wing-bow rich red. Secondaries bright bay. Flights black inner web, red outside edging. Remainder lustrous green-black.

Malay
Male and Female

Plumage, Female: Any shade of cinnamon with dark purple-tinted hackle; quite free of ticks, spangles, or pencilling, or white in tail and wings. Partridge-marked and Clay hens with golden hackle are also allowed.

Plumage, Pile Variety, Male: Hackle, saddle, back, and wing-bow rich red. Secondaries bright bay. Flights white inner web, red outside edging. Remainder cream-white.

Plumage, Female: Hackle gold. Breast salmon. Remainder cream-white.

Plumage, Spangled Variety, Male: Breast, underparts, thighs and tail an admixture of red and white. Remainder, each feather somewhat resembling tortoise-shell in the blending of red or chestnut with black, and with a bold white tip or spangle, the flight feathers and tail as tri-coloured as possible.

Plumage, Female: Rich dark red or chestnut boldly marked with black and white.

Plumage, White Variety, Male and Female: Pure white free from any yellow, black, or ruddy feathers.

In both sexes: Beak yellow or horn. Eyes pearl, white, yellow or daw with a green shade but the lighter the better; a red or foxy tinge very objectionable. Comb, face, throat, wattles and ear-lobes brilliant red. Legs and feet rich yellow, although in the Black a slight duskiness may be overlooked.

Note: The foregoing are the principal varieties, and others are not kept or bred in sufficient numbers to warrant description. The above colours and markings are ideal, but type and quality are the most important points in the Malay fowl.

Standard Weights: Male about 11 lb. Female about 9 lb.

SCALE OF POINTS

TYPE (SHOULDERS 7, CURVES AND CARRIAGE 16, REACH 12)	35
HEAD	16
EYES	9
LEGS	10
FEATHERING	10
COLOUR	6
TAIL	6
CONDITION	8
	100

Note: Size to be left to the discretion of the Judge.

Serious Defects: Any clear evidence of an alien cross. Lack of size. Single, spreading, or pea comb. Red eye, bow legs, knock knees, bad feet, in short, any defect not sufficiently penalised by the deduction of the maximum number of points allowed in the above scale.

MARANS

ORIGIN: French
CLASSIFICATION: Heavy
EGG COLOUR: Dark Brown

TAKING its name from the town of Marans in France, this breed has in its make-up such breeds as the Coucou de Malines, Croad Langshan, Rennes, Faverolle, Barred Rock, Braekel and Gatinaise. Imported into this country round about 1929, it has developed as a dual-purpose sitting breed. Like other barred breeds the Cuckoo Maran hens can be mated with males of other suitable unbarred breeds to give sexlinked offspring on the white head-spot distinguishing characteristic.

GENERAL CHARACTERISTICS: MALE

Carriage: Active, compact and graceful.

Type: Body of medium length with good width and depth throughout; front broad, full and deep. Breast long, well fleshed, of good width, and without keeliness. Tail well carried, high.

Head: Refined. Beak deep and of medium size. Eyes large and prominent; pupil large and defined. Comb single, medium size, straight, erect, with five to seven serrations, and of fine texture. Face smooth. Wattles of medium size and fine texture.

Neck: Of medium length and not too profusely feathered.

Legs and Feet: Legs of medium length, wide apart, and good quality bone. Thighs well fleshed, but not heavy in bone. Shanks clean and unfeathered. Toes, four, well spread and straight.

Plumage: Fairly tight and of silky texture generally.

Handling: Firm, as befits a table breed. Flesh white, and skin of fine texture.

FEMALE

General characteristics similar to those of the male, allowing for the natural sexual differences. Table and laying qualities to be taken carefully into account jointly.

COLOUR

Plumage, Black Variety, Male and Female: Black with a beetle-green sheen.

Plumage, Dark Cuckoo Variety, Male and Female: Cuckoo throughout, each feather barred across with bands of blue-black. A lighter shaded neck in both male and female, and also back in the male, is permissible if definitely barred. Cuckoo throughout is the ideal, as even as possible.

Plumage, Golden Cuckoo Variety, Male: Hackles bluish-grey with golden and black bars; neck paler than saddle. Breast bluish-grey with black bars, pale golden shading on upper part. Thighs and fluff light bluish-grey with medium black barring. Back, shoulders and wing bows bluish-grey with rich bright golden and black bars. Wing bars bluish-grey with black bars: golden fringe permissible. Wings, primaries dark blue-grey, lightly barred; secondaries dark blue-grey, lightly barred, with slight golden fringe. Tail dark blue-grey barred with black; coverts blue-grey barred with black. General Cuckoo markings.

Plumage, Female: Hackle medium bluish-grey with golden and black bars. Breast dark bluish-grey with black bars, pale golden shading on upper parts. Remainder dark bluish-grey with black bars. Cuckoo markings.

Plumage, Silver Cuckoo Variety, Male: Mainly white in neck and showing white on upper part of breast, also on top. Remainder barred throughout, with lighter ground colour than the Dark Cuckoo.

Plumage, Female: Mainly white in neck and showing white on upper part of breast. Remainder barred throughout, with lighter ground colour than the Dark Cuckoo.

In both sexes: Beak, white or horn. Eyes, red or bright orange preferred. Comb, face, wattles and ear-lobes, red. Legs and feet white.

Standard Weights: Cock 8 lb; cockerel 7 lb. Hen 7 lb; pullet 6 lb.

Marans
Dark Cuckoo Male
and Female

SCALE OF POINTS

TYPE, CARRIAGE AND TABLE MERITS (TO INCLUDE TYPE OF BREAST AND FLESHING, ALSO QUALITY OF FLESH)	40
SIZE AND QUALITY	20
COLOUR AND MARKINGS	15
HEAD	10
CONDITION	10
LEGS AND FEET	5
	100

Serious Defects: Feathered shanks. General coarseness. Lack of activity. Superfine bone. Any points against utility or reproductive values. **Defects** (for which a bird may be passed): Deformities, crooked breast bone, other than four toes, etc. **Minor Defects** (in Blacks): Restricted white in undercolour in both sexes. A little darkish pigmentation in white shanks.

MINORCA

ORIGIN: Mediterranean
CLASSIFICATION: Light
EGG COLOUR: White

THE Minorca has been developed in this country as our heaviest light breed, and was at one time famous for its extra-large white eggs. Crossing with the Langshan and other heavy breeds did not improve the egg production of the breed, and concentration on exaggerated headgear had a similar effect. Those times are passed and wiser counsels now prevail. The result is that a much better balanced type is aimed for on the show-bench, with moderate size of lobes and of comb, and a more prominent frontal.

GENERAL CHARACTERISTICS: MALE

Carriage: Upright and graceful.

Type: Body broad at the shoulders, square and compact. Breast full and rounded. Wings moderate in length, neat fitting close to body. Tail full, sickles long, well arched and carried well back.

Minorca
Black Male,
Blue Female

Head: Long and broad, so as correctly to carry the comb quite erect. Beak fairly long and stout. Eye full, bright and expressive. Comb single or rose. The single large, evenly serrated, perfectly upright, firmly set on the head, straight in front, free from any twist or thumb mark, reaching well to the back of the head, moderately rough in texture and free from any side sprigs. The rose oblong shape, broad at the base over the eyes, closely fitting, upright, firmly carried, full in front and tapering gradually to the " leader " at the back, surface evenly covered with small nodules or points, free from hollowness, the " leader " to follow the curve of the neck but not to touch the hackle. Face fine in quality, as free from feathers or hairs as possible, and not showing any white. Ear-lobes medium in size, almond shaped, smooth, flat, fitting close to the head. The lobe should not exceed $2\frac{3}{4}$ inches deep and $1\frac{1}{2}$ inches at its widest part on the top, tapering as the Valencia almond in shape. Wattles, long and rounded at the end.

Neck: Long, nicely arched, with flowing hackle.

Legs and Feet: Legs of medium length, and thighs stout. Toes four.

FEMALE

The general characteristics are similar to those of the male, allowing for the natural sexual differences, with the exception of the single comb which drops well down over the side of the face, so as not to obstruct the sight, and the ear-lobes which are $1\frac{3}{4}$ inches deep and $1\frac{1}{4}$ inch wide.

COLOUR

Plumage, Black Variety, Male and Female: Glossy black.
In both sexes: Beak dark horn. Eyes dark. Comb, face and wattles blood-red. Ear-lobes pure white. Legs black or very dark slate, the latter in adults only.

Plumage, White Variety, Male and Female: Glossy white.
In both sexes: Beak white. Eyes red. Comb, face and wattles blood-red. Legs pinky-white.

Plumage, Blue Variety, Male and Female: Soft medium blue, free from lacing, the male darker in hackles, wing bows and back.
In both sexes: Beak, comb, face, ear-lobes and wattles as in the black. Eyes dark brown (darker the better). Legs blue to slate.

Standard Weights: Cock 7 to 8 lb; cockerel 6 to 8 lb. Hen 6 to 8 lb; pullet 6 to 7 lb.

SCALE OF POINTS

Black and White Varieties

STYLE, SYMMETRY (TYPE)	10
SIZE	15
FACE	15
COMB	15
EAR-LOBES	10
LEGS, EYES AND BEAK	8
COLOUR	10
CONDITION	10
BREAST BONE	7
	100

Blue Variety

TYPE	15
SIZE	10
CONDITION	10
HEAD, COMB AND EAR-LOBES	5
COLOUR	0
	100

Serious Defects: White or blue in face. Feathers on legs. Other than four toes. Wry or squirrel tail. Plumage other than black, blue or white in the several varieties. Legs other than black or slate in the Black, blue or slate-blue in the Blue, or white in the White. Sidesprigs in comb.

MODERN GAME

ORIGIN: British
CLASSIFICATION: Heavy
EGG COLOUR: Tinted

By the introduction of Malay crosses, and with the skill of British fanciers, the Modern Game fowl was evolved. Black-reds, Duck-wings, Brown-reds, Piles, and Birchens are the recognised varieties, the general characteristics being the same for each.

GENERAL CHARACTERISTICS: MALE

Carriage: Upstanding and active.

Type: Body short, flat back, wide front and tapering to the tail, shaped like a smoothing iron. Shoulders prominent and carried well up. Wings short and strong. Tail short, fine, closely whipped together, carried slightly above the level of the body, the sickles narrow, well pointed and only slightly curved.

Head: Long, snaky and narrow between the eyes. Beak long, gracefully curved and strong at the base. Eyes prominent. Comb single, small, upright, of fine texture, evenly serrated. Face smooth. Ear-lobes and wattles fine and small to match the comb.

(It is customary to dub Game cocks, to remove comb, ear-lobes, and wattles, and thus leave the head and lower jaw smooth and free from ridges.)

Neck: Long and slightly arched, fitted with " wiry " feathers, but thin at the junction with the body.

Legs and Feet: Legs long and well rounded. Thighs muscular. Shanks free of feathers. Toes, four, long, fine and straight, the fourth (or hind) toe straight out and flat on the ground, not downwards against the ball of the foot (or " duck-footed "), which is most objectionable.

Plumage: Short and hard.

FEMALE

The general characteristics are similar to those of the male, allowing for the natural sexual differences.

COLOUR

Plumage, Birchen Variety, Male: Hackle, back, saddle, shoulder coverts and wing bows silver-white, the neck hackle with narrow black striping. Remainder rich black, the breast having a narrow silver margin around each feather, giving it a regular laced appearance gradually diminishing to perfect black thighs.

Plumage, Female: Hackle similar to that of the male. Remainder rich black, the breast very delicately laced as in the male.

In both sexes: Beak dark horn. Eyes black. Comb, face, wattles and ear-lobes dark purple. Legs and feet black.

Plumage, Black-red Variety, Male: Cap orange-red. Neck hackle light orange, free from black stripe. Back and saddle rich crimson. Wings: bow, orange ; bar green-black ; primaries black; secondaries rich bay on the outer edge, black on the inner and tips,

Modern Game
Pile Male,
Brown-red Female

the rich bay alone showing when the wing is closed. Remainder green-black.

Plumage, Female: Hackle gold, slightly striped with black, running to clear gold on the cap. Breast rich salmon, running to ash on thighs. Tail black, except the top feathers, which should match the body colour. Remainder light partridge-brown ground, very finely pencilled, and a slight golden tinge pervading the whole, which should be even throughout, free from any ruddiness whatever and with no trace of pencilling on the flight feathers.

In both sexes: Beak dark green. Eyes, comb, face, wattles and ear-lobes bright red. Legs willow.

Plumage, Brown-red Variety, Male: Hackle, back and wing bow bright lemon, the neck hackle feathers striped down the centre with green-black, not brown. Remainder green-black, the breast feathers edged with pale lemon as low as the top of the thighs.

Plumage, Female: Neck hackle light lemon to the top of the head, the lower feathers being striped with green-black. Remainder green-black, the breast laced as in the male, the shoulders free from ticking and the back from lacing. (*Note:* There should be only two colours in Brown-red Game, viz., lemon and black. In the male the lemon should be very rich and bright, and in the female light; the black in both sexes should have a bright green gloss known as beetle-green.)

In both sexes: Beak very dark horn, black preferred. Eyes, comb, face, wattles, ear-lobes, legs and feet black.

Plumage, Golden Duckwing Variety, Male: Hackle cream-white, free from striping. Back and saddle pale orange or rich yellow. Wings: bow, pale orange or rich yellow; bars and primaries, black with blue sheen ; secondaries pure white on the outer edge, black on inner and tips, the pure white alone showing when the wing is closed. Remainder black with blue sheen.

Plumage, Female: Hackle silver-white, finely striped with black. Breast salmon, diminishing to ash-grey on thighs. Tail black, except top feathers, which should match the body colour. Remainder French or steel-grey, very lightly pencilled with black, and even throughout.

In both sexes: Beak dark horn. Eyes ruby red. Comb, face, wattles and ear-lobes red. Legs and feet willow.

Plumage, Silver Duckwing Variety, Male: Hackle, back, saddle, shoulder coverts and wing bows silver-white. Secondaries pure white on the outer edge and black on the inner, with tips of bay, the white alone showing when the wing is closed. Remainder lustrous blue-black.

Plumage, Female: Hackle silver-white, finely striped with black. Breast pale salmon, diminishing to pale ash-grey on thighs. Tail black, except top feathers which should match the body colour. Remainder light French grey with almost invisible black pencilling.
In both sexes: Beak, etc., as in the Golden Variety.

Plumage, Pile Variety, Male: Hackle one shade of bright orange-yellow. (Dark or washy hackles to be avoided.) Back and saddle rich maroon. Wings: bow, maroon; bar white and free from splashes; primaries white; secondaries dark chestnut on the outer edge and white on the inner and tips, the dark chestnut alone showing when wing is closed. Remainder pure white.

Plumage, Female: Hackle white, tinged with gold. Breast rich salmon-red. Remainder pure white.
In both sexes: Beak yellow. Eyes bright cherry-red. Comb, face, wattles and ear-lobes red. Legs and feet rich orange-yellow.

Standard Weights: Male 7 to 9 lb. Female 5 to 7 lb.

SCALE OF POINTS

TYPE AND STYLE	30
HEAD AND NECK	10
EYES	10
TAIL	10
LEGS AND FEET	10
COLOUR	20
CONDITION AND SHORTNESS OF FEATHER.	10
	100

Serious Defects: Eyes other than standard colour. Flat shins. Crooked breast. Twisted toes or " duck " feet. Wry tail. Crooked back.

MODERN LANGSHAN

ORIGIN: Asiatic
CLASSIFICATION: Heavy
EGG COLOUR: Brown

THE Modern Langshan has been developed on different lines from the Croad, and is quite another type of bird. It is much longer in the shanks and of a totally different outline from the original Croad.

GENERAL CHARACTERISTICS: MALE

Carriage: Graceful, upright, alert, strong on the leg with the bearing of an active bird.

Type: Body long and broad but by no means deep. Back horizontal when in normal attitude, and with close compact plumage. Shoulders broad and abundantly furnished saddle. Wings large, closely carried, but neither " clipped " nor " pinched in ". Tail full, flowing, spread at base, carried fairly high but not " squirrel ", furnished with abundant side-hangers and two sickles, each feather tapering to a point.

Head: Fine. Beak fairly long and slightly curved. Eyes large. Comb single, straight, upright, fairly small, and evenly serrated with five or six spikes. Face of fine texture. Ear-lobes medium size, pendent and inclined to fold. Wattles medium length, fine texture, neatly rounded.

Neck: Fairly long, broad at base and covered with full hackle.

Legs and Feet: Legs rather long, strong and wide apart. Thighs covered with closely fitting feathers, especially around the hocks. Shanks strong but not coarse-boned, with an even fringe of feathers (not heavy) on the outer sides. Toes, four, long, straight and well spread, the outer toe (and that alone) slightly feathered.

Plumage: Close and smooth.

FEMALE

With the exception of the tail (not carried high) the general characteristics are similar to those of the male, allowing for the natural sexual differences.

COLOUR

Plumage, Black Variety, Male and Female: Black with a brilliant beetle-green sheen.

In both sexes: Beak dark horn to black. Eyes dark brown to black, the darker the better. Comb, face, wattles and ear-lobes brilliant red. Legs and feet dark grey with black scales in front and down the toes, showing pink between the scales (especially down the outer sides of the shanks) and on the skin between the toes. Toe-nails white. Under-foot pink-white. Skin of the body and thighs white and transparent.

Plumage, Blue Variety, Male: Hackles, back, tail, sickles, side-hangers, and wing bow, rich deep slate, the darker the better, with brilliant purple sheen. Remainder clear slate-blue, each feather distinctly laced (edged) with the same dark shade as the back, the contrast between delicate ground and dark lacing being well defined.

Plumage, Female: Head and upper part of neck rich dark slate. Remainder clear slate-blue, each feather distinctly laced (edged) with dark slate, the lacing being well defined.

In both sexes: Beak, etc., as in the Black.

Plumage, White Variety, Male and Female: Pure white, with brilliant silver gloss.

In both sexes: Beak white with a pink shade near the lower

Modern Langshan
Female

edges. Legs and feet light grey or slate showing pink between the scales and on the skin between the toes. Eyes, comb, face, wattles, ear-lobes, toe-nails, under-foot and skin as in the Black.

Standard Weights: Cock 10 lb; cockerel 8 lb. Hen 8 lb; pullet 6 lb.

SCALE OF POINTS

TYPE AND CARRIAGE	35
HEAD	15
LEGS AND FEET	10
SIZE	10
COLOUR AND PLUMAGE	20
CONDITION	10
	100

Serious Defects: Yellow skin. Yellow base of beak. Yellow or orange-coloured eyes. Yellow around the eyes. Yellow shanks or under-foot. Legs other than standard colour. Shanks not feathered. More than four toes. Permanent white in the face or ear-lobes. Comb with side sprigs, or other than single. Wry or squirrel tail. Coloured feathers.

Defects: Absence of pink between toes. Feathering on middle toes. Outer toe not feathered. Too scantily or too heavily feathered shanks or outer toes. Twisted toes. Short shanks. Crooked breast. Twisted or falling comb. General coarseness. Too much fluff. Purple sheen (in Black); yellow shade (in White).

NEW HAMPSHIRE RED

ORIGIN: American
CLASSIFICATION: Heavy
EGG COLOUR: Tinted to brown

As if to copy the farmers in the State of Rhode Island who developed the breed carrying its name, those in the neighbouring State of New Hampshire developed and named their breed. The New Hampshire Red was bred by selection from the Rhode Island

New Hampshire Red
Male and Female

Red without the introduction of any other breed, taking some thirty years to reach Standardisation in 1935. Early maturity, quick feathering, and a plump carcase are particular features of the breed. Its body-shape and colouring are very different from those of the Rhode Island Red.

GENERAL CHARACTERISTICS: MALE

Carriage: Active and well balanced.

Type: Body of medium length, relatively broad, deep and well rounded. Back of medium length, broad for entire length, gradual concave sweep to tail. Breast deep, full, broad and well rounded, the keel relatively long and extending well to the front at the breast. Wings moderately large, well folded, carried horizontally and close to the body, fronts well covered by breast feathers; primaries and secondaries broad and overlapping in natural order when wing is folded. Tail of medium length, well spread and carried at angle of 45 degrees. Sickles medium in length, extending well beyond the main tail; lesser sickles and coverts medium length and broad; main tail feathers broad and overlapping.

Head: Of medium length, fairly deep, inclined to be flat on top rather than round. Beak strong, medium length, regularly curved. Eyes large, full and prominent, moderately high in the head. Comb single, medium size, well developed, set firmly on head, perfectly straight and upright, having five well defined points, those in front and rear smaller than those in centre; the blade smooth and inclining slightly downward, not following too closely the shape of the neck. Face smooth, full in front of the eyes; skin of fine texture. Wattles moderately large, uniform, free from folds or wrinkles. Ear-lobes elongated-oval, smooth and close to head.

Neck: Of medium length, well arched, hackle abundant, flowing well over shoulders. Moderately close feathered.

Legs and Feet: Legs well apart, straight when viewed from the front. Lower thighs large, muscular and of medium length. Toes, four, medium length, straight and well spread.

Plumage: Feather character to be of broad firm structure, overlapping well and fitting tightly to the body. Fluff moderately full.

FEMALE

The general characteristics are similar to those of the male, allowing for the natural sexual differences. Comb slightly tilted at rear.

114

Wattles of medium size, well developed and well rounded. Tail moderately well spread, carried at angle of 35 degrees. Wings rather large carried nearly horizontal.

COLOUR

Plumage, Male: Head brilliant reddish bay. Breast and neck medium chestnut red. Back brilliant deep chestnut red. Saddle rich brilliant reddish bay, slightly darker than neck. Wing fronts medium chestnut red, bows brilliant deep chestnut red; coverts deep-chestnut red; primaries, upper web medium red, lower web black edged with medium red; primary coverts black edged with medium red; secondaries, upper web medium chestnut red having broad stripe extending along shaft to within inch of tip, lower web medium chestnut red; shaft red. Tail, main feathers black; sickles rich lustrous greenish-black; coverts lustrous greenish-black edged with deep chestnut red; lesser coverts deep chestnut red. Body and fluff medium chestnut red. Lower thighs medium chestnut red. Undercolour in all sections light salmon, a slight smoky tinge not a defect.

Plumage, Female: Head medium chestnut red. Neck medium chestnut red, each feather edged with brilliant chestnut red; lower neck feathers distinctly tipped with black; feathers in front of neck medium chestnut red. Wing fronts, bows and coverts medium chestnut red; primaries, upper web medium red, lower web medium red with narrow stripe of black extending along shaft; shaft medium red; primary coverts black edged with medium red; secondaries, lower web medium chestnut red, upper web medium chestnut red with black marking extending along edge of shaft two-thirds its length. Back, breast, lower thighs, body and fluff medium chestnut red. Tail, main feathers black edged with medium chestnut red; shaft medium chestnut red. Undercolour as in male.

In both sexes: Beak reddish brown. Eyes bay. Comb, face wattles and ear-lobes bright red. Legs and toes rich yellow, tinged with reddish horn. Line of reddish pigment down sides of shanks extending to tips of toes desirable in male.

Standard Weights: Cock 8½ lb; cockerel 7½ lb. Hen 6½ lb; pullet 5½ lb.

SCALE OF POINTS

TYPE AND CARRIAGE	25
COLOUR	20
DUAL-PURPOSE QUALITY	15

Continued on next page

115

New Hampshire Red Points—continued

    HEAD.. 10
    SIZE AND SYMMETRY...................... 10
    LEGS AND FEET......................... 10
    CONDITION.............................. 10
                                              ——
                                             100
                                              ══

NORTH HOLLAND BLUE

ORIGIN: Dutch
CLASSIFICATION: Heavy
EGG COLOUR: Tinted

THE blood of the Malines in the make-up of this breed is seen in its quick maturity and rapid growth. To further its commercial importance its utility properties are valued on the show bench in preference to markings. Lightly feathered shanks are a standardised characteristic, and the male is lighter in colour than the female. As a barred breed, the females when mated with unbarred males of breeds with dark downs, produce sex-linked offspring, the cockerel chicks having the white head-spot when hatched, which is absent in the pullet chicks.

GENERAL CHARACTERISTICS: MALE

Carriage: Upright, bold and alert.

Type: Body substantial in build, yet compact. Back broad, flat, horizontal, reasonably long, with slight rise to tail, broad saddle, prominent shoulders. Breast full, rounded and prominent; breast bone long, well fleshed and rounded-off, neither shallow nor keely. Well rounded sides, good depth of body, with a well-developed abdomen. Wings strong, well developed and close to the body. Tail broad, short, well spread with medium furnishings.

Head: Rounded and of medium length. Beak stout and short. Eyes full, bold, with keen expression, the pupils well formed and large. Face smooth, full, fine texture and without heavy eyebrows. Comb single, upright, of medium size and fine texture, with five to seven neat even serrations, following slightly the curve of the neck at the back. Ear-lobes of medium size and silky. Wattles medium in length and size, and fine in texture.

116

North Holland Blue
Male and Female

Neck: Somewhat broad, medium in length and not too profusely feathered.

Legs and Feet: Legs of medium length, wide apart, well formed and good quality bone with close scales. Thighs well developed and fleshed. Toes, four, straight, and well spread. Shanks lightly feathered, including outer toe.

Plumage: In general not too profuse. Fine in texture.

Handling: Firm, as befits a table breed. Skin thin and of fine texture. Flesh high quality table grade.

FEMALE

With the exception of the shanks, which are more heavily feathered, the general characteristics are similar to those of the male, allowing for natural sexual differences.

COLOUR

Plumage, Male: Lighter blue-grey than female, barred. Undercolour very pale blue-grey, barring immaterial.

Plumage, Female: Dark grey-blue, slightly barred. Undercolour paler blue-grey, barring immaterial.

In both sexes: Beak white superimposed blue. Eyes orange to red. Comb, wattles, face and ear-lobes red. Shanks white, but may be shaded with blue. Skin white.

Standard Weights: Cock $8\frac{1}{2}$ to $10\frac{1}{2}$ lb; cockerel $7\frac{1}{2}$ to $9\frac{1}{2}$ lb. Hen 7 to 9 lb; pullet 6 to 8 lb.

SCALE OF POINTS

TYPE AND CARRIAGE	10
INDICATION OF TABLE MERITS	20
INDICATION OF EGG PRODUCTION MERITS	20
COLOUR AND MARKINGS	20
HEAD	10
LEGS AND FEET	10
CONDITION	10
	100

Defects : General coarseness. Superfine bone. Unfeathered shanks. Any points against table, laying or reproductive qualities. Birds may be passed for deformities, crooked breast bone, serious defects, and clean shanks devoid of any feathering.

OLD ENGLISH GAME

WHEN the Romans invaded Britain, Julius Caesar wrote in his commentaries that the Britons kept fowls for pleasure and diversion but not for table purposes. Many well-known authorities have considered that cock-fighting was the diversion. In 1849 an Act of Parliament was passed making cock-fighting illegal in this country, and with poultry exhibitions then taking root, many breeders began to exhibit Game fowls. Over thirty colours of Old English Game have been known.

GENERAL CHARACTERISTICS: MALE

Carriage: Bold, sprightly, the movement quick and graceful, as if ready for any emergency.

Type: Back short and flat, broad at the shoulders, and tapering to the tail. Breast broad, full, and prominent with large pectoral muscles, and breast bone not deep or pointed. Wings large, long and powerful with large strong quills, amply protecting the thighs. Belly small and tight. Tail large, carried upwards and spread, main feathers and quills large and strong.

Head: (It is customary to dub O.E.G. cockerels.) Small and tapering. Beak big, boxing (i.e., the upper mandible shutting tightly and closely over the lower one), crooked or hawk-like, pointed, strong at the base. Eyes large, bold and full of expression. Comb single, small, thin, upright, of fine texture in undubbed chickens and hens. Face and throat skin flexible and loose. Ear-lobes and wattles fine, small and thin.

Neck: Large boned, round, of fair length, and very strong at the junction with the body, furnished with long and wiry feathers covering the shoulders.

Legs and Feet: Legs strong. Thighs short, round and muscular, following the line of the body, or slightly curved. Shanks strong, clean boned, sinewy, close scaled (not flat and " gummy "), not stiffly upright or too wide apart, and with a good bend or angle at the hock, fitted with hard and fine spurs set low. Toes, four,

thin, straight and tapering, terminating in long, strong, curved nails, the fourth (or hind) toe strong, straight out and flat on the ground.
Handling: Well balanced, hard yet light fleshed, " corky ", with plenty of muscle, and strong contraction of the wings and thighs to the body.
Plumage: Hard and glossy, without much fluff.

FEMALE

With the exception of the tail, which is inclined to fan-shape and carried well up, the general characteristics are similar to those of the male, allowing for the natural sexual differences.

COLOUR

The following colours are recognised by the Oxford O.E.G. Fowl Club:—

Black Breasted Black-red Male: Hackle, shoulders and saddle rich dark red (the colour of the shoulders of a black breasted red). The rest of the plumage black.

Female: Body brown, mixed with umber brown, hackle striped red, breast red-brown, tail and primary wing feathers black.

In both sexes: Fluff (i.e., the down at the roots of the feathers next to the skin) black. Eyes, beak, legs and nails black. Face gipsy or purple.

Black Breasted Red Male: Breast, thighs, belly and tail black; wing bars steel blue, secondaries (when closed) bay; hackle and saddle feathers orange-red; shoulders deep crimson-scarlet.

Female: Hackle golden, lightly striped with black; breast robin; belly ash-grey; back, shoulders and wings a good even partridge, primaries dark, also tail.

In both sexes the dark-legged birds should have grey fluff, the white and yellow legged, white fluff. Face scarlet-red. Legs willow, yellow, white carp or olive.

Shady or Streaky Breasted Light-red Male: Hackle and back a shade lighter than the Black Breasted Red male and sometimes red wing bars.

Female: Wheaten, a pale cream colour (like wheat) with clear red hackle; tail and primaries nearly black. The Red Wheaten (the colour of red wheat), or light brick red in body and wings; hackle dark red; tail dark.

In both sexes: Fluff white. Legs white or yellow.

Black Breasted Silver Duckwing Male: Resembles the black breasted red in his black markings and blue wing bars; rest of the plumage clear silvery white.

Old English Game
Brown-red Male,
Blue Dun Female

Female: Hackle white, lightly striped black; body and wings even silvery grey; breast pale salmon; primaries and tail nearly black.

In both sexes: Fluff light grey. Face red, eyes pearl. Legs and beak white. Or eyes red and legs dark.

Black Breasted Yellow Duckwing Male: Hackle and saddle yellow straw; shoulders deep golden; wing bars steel blue; secondaries white when closed; rest of plumage black.

Female: Breast deeper, richer colour and body slightly browner tinge than the silver female.

In both sexes: Fluff light grey. Face red. Legs yellow, willow or dark.

Black Breasted Birchen Duckwing Male: Hackle deep rich straw, may be lightly striped; shoulders maroon; otherwise same as preceding.

Female: Shade darker than Yellow Duckwing female; hackle more heavily striped with black, and often foxy on the shoulders.

In both sexes: Face slightly darker than in Yellow Duckwing. Legs yellow or dark.

Black Breasted Dark Grey Male: Like the Black Breasted Red, except hackle, saddle and shoulders a dark silver-grey, often striped with black.

Female: Nearly black, with grey striped hackle, or body very dark grey.

In both sexes: Fluff black. Beak, eyes and legs black. Face gipsy or purple.

Other Greys may have laced, streaked or mottled grey or throstle breasts; hackle, saddle and shoulders more or less striped with black; legs and eyes dark; the females dark grey to match; fluff in both sexes, light or dark grey.

Note: Greys all differ from Duckwings in having the secondaries, when closed, black; or, if grey, wanting the steel blue bar across them.

Clear Mealy Breasted Mealy Grey Male: Nearly white breasted, with hackle and saddle the same, lightly striped; plumage and most of the tail grey.

Female: Light grey.

In both sexes: Fluff light grey. Eyes and legs dark.

Brown Breasted Brown-red Male: Breast, thighs, belly and closed wing mahogany brown; hackle and saddle almost similar; shoulders crimson; primaries and tail black or dark bronze-brown.

Female: Dark mottled brown with light shafts to the feathers.

Old English Game
Tassel Pile Male,
Grey Female

In both sexes: Fluff black. Face deep crimson or purple. Eyes and legs dark.

Streaky Breasted Orange-red Male: Breast streaked, laced or pheasant, black, marked with brown or copper colour; hackle and saddle brassy or coppery orange colour; shoulders crimson; the rest of the wings and the tail black.

Female: Black or nearly black body, with tinsel hackle striped with black, or dark mottled brown and gold striped hackle.

In both sexes: Fluff black or nearly so. Face, eyes and legs dark.

Ginger Breasted Ginger-red Male: Breast and thighs deep yellow ochre, either clear or slightly pencilled or spotted; hackle and saddle red golden; shoulders crimson-red; tail and flight feathers bronzy.

Female: Golden yellow throughout, pencilled or spangled, particularly on back and wings, with bronze; tail pencilled bronze or dark.

In both sexes: Fluff dark. Beak, legs and eyes dark or yellow. Face purple or crimson.

Dun Breasted Blue Dun Male: Breast, belly, thighs, tail and closed secondaries the colour of a new slate, sometimes the breast marked with the same colour two shades darker; hackle, saddle and shoulders, and sometimes the tail coverts and the primaries two shades darker (like a slate colour after being wetted).

Female: Blue slate colour, with dark hackle like the male, often marked or laced all over with the darker shade.

In both sexes: Fluff slate blue. Eyes, face and legs dark.

Streaky Breasted Red Dun Male: Breast slatey, streaked with copper red; hackle and saddle striped with slate or dark striped; shoulders crimson; wing bars and closed secondaries slate, or marked a little with brown; tail slatey or dark blue.

Female: Body slatey all over, or laced in a darker shade; hackle golden striped, and sometimes marked with gold on the breast.

In both sexes: Fluff dark slate. Legs dark or yellow.

Yellow, Silver and Honey Duns, Male and Female: These are coloured respectively with the following colours; the colour of new honeycomb is intended to describe the Honey Dun. They may have yellow or dark legs according to body colour, and white legs are permissible in the Silver Dun, as well as other coloured legs. The females are blue bodied with hackles to match their males. Smoky Duns are of a dull smoke colour throughout; legs and eyes should be dark.

Piles, Male: The Smock Breasted Blood Wing Pile is marked exactly like the Black Breasted Light Red, except that the black and the blue wing bars are exchanged for a clear cream-white. The breast may be streaked with red in Red Pile.

Female: White, with salmon breast and golden striped hackle, or streaked all over lightly with red.

In both sexes: Face and eyes red. Legs white, yellow or willow. *Note:* Other varieties of Piles may be streaky, marbled or robin breasted; and light lemon or custard in top colour, or Dun Piles having slate-blue markings in place of red. All Piles have white fluff.

Spangles, Male and Female: These have white tips to their feathers. The more of these spots and the more regularly they are distributed the better. The male should show white ends to the feathers on hackle and saddle. The ground colour may be red, black or brown, or a mixture of all three. Underfluff white; eyes and face red; legs any colour or mottled to match plumage.

White, Male and Female: This variety should be free from any coloured feathers. Fluff pure white. Beak and legs white. Face red; eyes pearl; or yellow legs and red eyes.

Black, Male and Female: This variety should be free from any white or coloured feathers and should possess dark legs, beaks, faces and black eyes, though red faces and red eyes are allowed at present; fluff black.

Furnesses, Brassy Backs and Polecats, Male and Female : These are Blacks with brass colour on their wings or back, and occasionally have yellow legs, which are allowed. The females are chiefly black, but often much streaked with grey-brown on breast and wings. Polecats are streaked with dark tan colour on hackles and saddle in the males. Legs dark.

Cuckoo, Male and Female: Cuckoo Breasted Cuckoo resembles the Plymouth Rock fowl in markings of a blue-grey barred plumage. Faces and eyes red; legs various.

Variations of this colour are Yellow Cuckoos, also Creels, Creoles, Cirches, Mackerels in different provincial dialects, having some mixture of gold or red in the plumage, often extremely pretty. Legs white or yellow. Fluff white.

Brown Breasted Yellow Birchen, Male: Breast reddish brown; hackles and saddle straw, striped birchen brown; shoulders old gold or birchen; wing bar and closed secondaries brown; tail brown or bronze-black.

Female: Yellow-brown, with grey hackle and robin breast.

In both sexes: Fluff light grey. Beak, legs and eyes yellow.

Hennies, Male and Female: Hencocks should in their plumage resemble hens as closely as possible. They should have their hackle and saddle feathers rounded and the tail coverts hen-like, and not have much sheen on their feathers. This breed often runs large and reachy, which is one of its characteristics. The two centre tail feathers should be straight.

Muffs and Tassels, Male and Female: Both Muffs and Tassels, or Topins, are recognised by the Club, there being famous strains of both, though now scarce. Tassels vary from a few long feathers (or lark tops) behind the comb to a good-sized bunch. They also occur in some strains of Hennies. Muffs of the old breed are stronger, heavier boned birds than the males bred today, and are rather loose in feather.

Notes:

(1) It is desirable that the toe-nails should match the legs and beak in colour in all Game Fowl, and that legs, eyes, beak and face match the male in all Game females.

(2) White or yellow legged birds may have white feathers in wings and tail.

(3) The fancier, when he speaks of a brown-red, means the streaky breasted orange-red; and when talking of Black-reds, intends one to infer a black breasted Light-red; while Black Breasted Dark Greys are erroneously called " Birchens ", although they have no birchen colour in them.

Standard Weights: Male 5½ lb to 5 lb 10 oz. (It is not considered desirable to breed males over 6 lb.) Female 4 to 5 lb.

SCALE OF POINTS

BODY (INCLUDING BREAST, BACK AND BELLY)	20
HANDLING (SYMMETRY, CLEVERNESS, HARDNESS OF FLESH AND FEATHERS, CONDITION AND CONSTITUTION)	15
HEAD (INCLUDING BEAK AND EYES)	10
NECK	6
SHANKS, SPURS AND FEET	10
PLUMAGE AND COLOUR	9
THIGHS	8
WINGS	7
TAIL	6
CARRIAGE, ACTION AND ACTIVITY	9
	100

Serious Defects: Thin thighs or neck. Flat sided. Deep keel. Pointed, crooked or indented breast bone. Thick insteps or toes. Duck feet. Straight or stork legs. In-knees. Soft flesh. Broken, soft or rotten plumage. Bad carriage or action. Any indication of weakness of constitution.

OLD ENGLISH PHEASANT FOWL

ORIGIN: British
CLASSIFICATION: Light
EGG COLOUR: White

THIS breed was given its name of Old English Pheasant Fowl about 1914, previous to which it had been called the Yorkshire Pheasant, Golden Pheasant and also the Old-fashioned Pheasant. That it is a very old English breed is certain. Some Northern breeders retained their strains as Yorkshire Pheasant Fowls until the present tag of " Old English " was brought officially into use. It has a meaty breast for a light breed, and has always been popular with farmers.

GENERAL CHARACTERISTICS: MALE

Carriage: Alert and active.

Type: Body rather long, deep and round with prominent shoulders. Tail flowing and set well back.

Head: Fine. Beak of medium size. Eyes bright and prominent. Comb rose, moderate in size, not impeding either sight or breathing, fine texture, evenly set, rather square front, the top flat and with plenty of work, tapering to a single leader (or spike) at the back, which should gracefully curve downwards, following the neck line but quite free from it. Face and wattles smooth, free from coarseness or wrinkles. Ear-lobes medium size, oval or almond shape, smooth.

Neck: Graceful.

Legs and Feet: Legs of medium length, well apart, neither coarse nor too fine. Shanks free from feathers. Toes, four, well spread.

FEMALE

With the exception of the tail (moderately whipped) the general characteristics are similar to those of the male, allowing for the natural sexual differences.

COLOUR

Plumage, Gold Variety, Male: Ground colour bright rich bay; back rich mahogany red; lacing, bars, striping, tipping and tail, beetle-green black. Hackles striped and slightly tipped; saddle a slightly deeper shade than neck. Breast laced. Wing bars (two) marked.

Plumage, Female: Ground colour bright rich bay; striping, tipping, spangling and tail beetle-green black. Neck with heavy stripe down centre of each feather. Wing bars crescent spangling; even and well-marked bars a point of great beauty. Tail with slight edging of ground colour carried up from the base along the upper edge of the tail. Remainder, each feather tipped with a crescent-shaped spangle. Shafts of all feathers, bay.

Plumage, Silver Variety, Male and Female: White with beetle-green black markings.

In both sexes: Beak horn. Eyes fiery red. Comb bright rich red. Face and wattles red. Ear-lobes white. Legs and feet slate-blue.

Standard Weights: Cock 6 to 7 lb; cockerel $5\frac{1}{2}$ to 6 lb. Hen 5 to 6 lb; pullet $4\frac{1}{2}$ to 5 lb.

SCALE OF POINTS

TYPE (INCLUDING LEGS)	20
HEAD (COMB 15, LOBES 5, OTHER POINTS 5)	25
MARKINGS	20
GROUND COLOUR OF BODY	15
PLUMAGE AND FLOW OF FEATHER	10
SIZE	5
CONDITION	5
	100

Serious Defects: Comb single or over either side. Blushed lobes. Sooty hackles. Definitely black breast in male. Superfine bone. Lack of size. Squirrel tail. Any deformity. Any other defects which would affect health, hardiness, productivity or activity up to 20 points.

Old English Pheasant
Fowl
Male and Female

ORPINGTON

ORIGIN: British
CLASSIFICATION: Heavy
EGG COLOUR: Brown

IN the Orpington we have an English breed named after the village in Kent where the originator, William Cook, had his farm. He introduced the Black variety in 1886, the White in 1889, and the Buff in 1894. Within five years of the original Black Orpington being introduced exhibition breeders were crossing Langshan and Cochin and exhibiting the offspring as Black Orpingtons, the birds fetching high prices, and attracting many for their immense size. But this crossing at once turned a dual-purpose breed into one solely for show purposes, and it has remained so till today. A late introduction, the Jubilee Orpington, is now never seen.

GENERAL CHARACTERISTICS: MALE

Carriage: Bold, upright and graceful; that of an active fowl.

Type: Body deep, broad and cobby. Back nicely curved with a somewhat short, concave outline. Saddle wide and slightly rising, with full hackle. Breast broad, deep and well rounded, not flat. Wings small, nicely formed and carried closely to the body, the ends almost hidden by the saddle hackle. Tail rather short, compact, flowing and high, but by no means a squirrel tail.

Head: Small and neat, fairly full over the eyes. Beak strong and nicely curved. Eyes large and bold. Comb single, small, firmly set on head, evenly serrated and free from side sprigs. In the Black variety comb may be single or rose, the latter, small, straight and firm, full of fine work or small spikes, level on top (not hollow in centre), narrowing behind to a distinct peak lying well down to the head (not sticking up). Face smooth. Wattles of medium length, rather oblong and nicely rounded at the bottom. Ear-lobes small and elongated.

130

Orpington
Buff Male and
Female

Neck: Of medium length, curved, compact and with full hackle.

Legs and Feet: Legs short and strong, the thighs almost hidden by the body feathers, well set apart. Toes, four, straight and well spread.

Plumage: Fairly profuse but close, not soft, loose and fluffy as in the Cochin, or close and hard as in the Game Fowl.

Handling: Firm.

FEMALE

The general characteristics are similar to those of the male. Her cushion should be wide but almost flat, and slightly rising to the tail, sufficient to give the back a graceful appearance with an outline approaching concave.

COLOUR

Plumage, Black Variety, Male and Female: Black with a green sheen.
In both sexes beak, etc., as in Blue. Soles of feet, white.

Plumage, Blue Variety, Male: Hackles, saddle, wing bow, back and tail dark slate-blue. Remainder medium slate-blue, each feather to show lacing of darker shade as on back.

Plumage, Female: Medium slate-blue, laced with darker shade all through, except head and neck, dark slate-blue.
In both sexes: Beak black. Eyes black or very dark brown, black preferred. Comb, face, wattles and ear-lobes bright red. Legs and feet black or blue. Toe-nails white.

Plumage, Buff Variety, Male and Female: Clear, even buff through-out to the skin.
In both sexes: Beak white or horn. Eyes red or brown. Comb, face, ear-lobes and wattles bright red. Legs, feet and toe-nails white. Skin white.

Plumage, White Variety, Male and Female: Pure snow white.
In both sexes: Beak, legs, feet and skin white. Eyes, face, ear-lobes and wattles red.

Standard Weights: Blue: Male 10 to 14 lb when fully matured; female $7\frac{1}{2}$ to $10\frac{1}{2}$ lb. Black: Male 10 lb; female 8 lb. Buff and White: Matured cockerels 8 to 10 lb; females 6 to 8 lb; old birds sometimes a little heavier in each case.

Orpington
Black Male,
White Female

SCALE OF POINTS

Blue Variety

TYPE	25
SIZE (WITH UTILITY QUALITIES)	20
HEAD	10
LEGS AND FEET	10
COLOUR AND PLUMAGE	25
CONDITION	10
	100

Buff and White Varieties

TYPE	30
SIZE	10
HEAD	15
LEGS AND FEET	10
COLOUR	20
CONDITION	15
	100

Black Variety

TYPE (SHAPE) (BODY 15, BREAST 10, SADDLE 5)	30
SIZE	10
CARRIAGE	10
HEAD (SKULL 5, COMB 7, FACE 5, EYES 5, BEAK 3)	25
SKIN	5
LEGS AND FEET	5
PLUMAGE AND CONDITION	10
TAIL	5
	100

Serious Defects: Side spikes on comb. White in ear-lobes. Feather or fluff on shanks or feet. Long legs. Any deformity. Yellow skin or yellow on the shanks or feet of any variety. Any yellow or sappiness in the White. Coarseness in head, legs or feathers of the Buff.

Disqualifications: Trimming or faking.

PLYMOUTH ROCK

ORIGIN: American
CLASSIFICATION: Heavy
EGG COLOUR: Tinted

SPECIMENS of the Barred Plymouth Rock were first exhibited in America in 1869, and stock reached here in 1871. The White and Black varieties came as sports. About 1890 the Buff was exhibited in America and in England. The Barred Rock came to us as a dual-purpose breed, but it was developed to an exhibition ideal in which body size and frontal development were neglected in order to secure long narrow finely barred feathers. With the introduction of sex-linkage between Black Leghorn and Barred Rock for commercial purposes, utility breeders made use of the Canadian Barred Rock, a bird with roomy body, full breast, lower on the leg but coarser in barring.

GENERAL CHARACTERISTICS: MALE

Carriage: Alert, upright with bold appearance, well balanced and free from stiltiness.

Type: Body large, deep and compact, evenly balanced and symmetrical, broad, the keel bone long and straight. Back broad and of medium length, saddle hackle of good length and abundant. Breast broad and well rounded. Wings of medium size, carried well up, bow and tip covered by breast and saddle feathers respectively; flights carried horizontally. Tail medium size, rising slightly from the saddle to be carried neatly and not to be fan, squirrel, or wry tail, sickles medium length and nicely curved, coverts sufficiently abundant to cover the stiff feathers.

Head: Of medium size, strong and carried well up. Beak short, stout and slightly curved. Eyes large, bright and prominent. Comb single, medium in size, straight and erect with well-defined serrations, smooth and of fine texture, free from side sprigs and thumb marks. Face smooth. Ear-lobes well developed, pendent, and of fine texture. Wattles moderately rounded and of equal length, to correspond with size of comb, smooth and of fine texture.

Neck: Of medium length, slightly curved, a full hackle flowing over the shoulders.

Legs and Feet: Legs wide apart. Thighs large and of medium length. Shanks medium length, stout, well rounded, smooth and free from feathers. Toes, four, strong and perfectly straight, well spread and of medium length.

Fluff: Moderately full, carried closely to the body and of good texture.

Skin: Silky and fine in texture.

FEMALE

The general characteristics are similar to those of the male, allowing for the natural sexual differences, except that comb, ear-lobes and wattles are smaller, the neck is of medium length, carried slightly forward, and the tail is small and compact, carried well back.

COLOUR

Plumage, Barred Variety, Male and Female: Ground colour, white with bluish tinge, barred with black of a beetle-green sheen, the bars to be straight, moderately narrow, of equal breadth and sharply defined, to continue through the shafts of the feathers. Every feather to finish with a black tip. The fluff, or under colour, to be also barred. The neck and saddle hackles, wing bow and tail to correspond with the rest of the body, presenting a uniformity of colour throughout.

Plumage, Black Variety, Male and Female: Black with a beetle-green sheen.

Plumage, Buff Variety, Male and Female: Clear, sound, even golden buff throughout to the skin. The tail clear buff to harmonise with the body colour, and the undercolour (or fluff) and the quill of the feather also to harmonise with the surface colour. The male of more brilliant lustre than the female.

Plumage, Columbian Variety, Male and Female: Colour and markings as in Light Sussex.

Plumage, White Variety, Male and Female: Pure snow-white, any straw tinge to be avoided.

In both sexes: Beak yellow. Eyes rich bay. Face, comb, ear-lobes and wattles red. Legs and feet yellow.

Standard Weights: Buff Variety—Cock 8½ to 10½ lb; Cockerel 7 to 8½ lb. Hen 6 to 7½ lb; pullet 5 to 6½ lb. Other Varieties—Cock 9 to 10 lb; cockerel 7½ to 9½ lb. Hen 7 to 8 lb; pullet 6 to 7 lb.

Plymouth Rock
Buff Male and
Female

SCALE OF POINTS

Barred Variety

TYPE	30
BARRING	20
COLOUR	15
SIZE	10
LEGS AND FEET	10
CONDITION	10
HEAD	5
	100

Buff Variety

TYPE (SYMMETRY), SHAPE, SIZE AND CARRIAGE	30
COLOUR (GENERAL)	20
QUALITY AND TEXTURE (GENERAL)	15
CONDITION AND FITNESS	15
HEAD AND COMB	10
EYE COLOUR	5
LEGS AND FEET	5
	100

Other Varieties

TYPE	30
COLOUR	30
HEAD AND EYES	10
LEGS AND FEET	10
CONDITION	10
QUALITY AND TEXTURE	10
	100

Serious Defects: The slightest fluff or feather on shanks or feet, or unmistakable signs of feathers having been plucked from them. Legs other than yellow. White in ear-lobes. In the Barred, any feathers of any colour foreign to the variety, black feathers excepted; also lopped or rose comb, decidedly wry tail; crooked back, more than four toes, and entire absence of main tail feathers. Other than black feathers in the Black. Mealiness, or any black or white in wing or white in tail, spotted hackle, and in the male a spotted saddle and in the female a spotted cushion in the Buff. Any coloured feathers in the White.

Plymouth Rock
Barred Male and
Female

Disqualifications: Trimming, faking and any bodily deformity, split wing, slipped wing and non-growth of secondaries.

POLAND (Polish)

ORIGIN: Polish
CLASSIFICATION: Light
EGG COLOUR: White

THAT the Poland is a very old breed goes without saying, although its ancestry is none too clear. Many connect it with the breed named the Paduan or Patavinian fowl, although this original example is illustrated without muff or beard. Polish (Gold or Silver Spangled, Black or White) had a classification at the first poultry show in London in 1845, and was Standardised in the first Book of Standards of 1865, with White-crested Black, Golden and Silver varieties included. The White-crested (Black or Blue) varieties are without muffling, while the others have muffs.

GENERAL CHARACTERISTICS: MALE

Carriage: Sprightly and erect.

Type: Back fairly long, flat and tapering to the tail. Breast full and round. Flanks deep. Shoulders wide. Wings large and closely carried. Tail full, neatly spread, and carried somewhat low, not perpendicularly, the sickles and coverts abundant and well curved.

Head: Large, with a decidedly pronounced protuberance on top, and crested. Crest large, full, circular on top and free of any split or parting, high and smooth in front and compact in the centre, falling evenly with long untwisted or reverse-faced feathers far down the nape of the neck, and composed of feathers similar to those of the hackles. Beak of medium length, and having large nostrils rising above the curved line of the beak. Eyes large and full. Comb of horn type and very small if any (preference should be given to birds without a comb). Face smooth, without muffling in the White-crested Black or Blue varieties, and completely covered by muffling in the others.   Muffling large, full and compact, fitting around to the back of the eyes and almost hiding the face.   Ear-lobes very small and round, quite invisible in the muffled varieties. Wattles rather large and long in the White-crested varieties; the others are without wattles.

Neck: Long, with abundant hackle coming well over the shoulders.

Legs and Feet: Legs slender and fairly long, the shanks free of feathers.   Toes, four, well spread.

FEMALE

With the exception of the crest, which is of globular shape, the general characteristics are similar to those of the male, allowing for the natural sexual differences.

COLOUR

Plumage, Chamois or White Laced Buff Variety, Male: Buff ground with white lacing.   Crest white at the roots and tips, and as free as possible from whole white feathers.   Muffling mottled or laced, not solid buff.   Hackle tipped.   Wing-bar and secondaries laced, and primaries tipped.   Tail, sickles and coverts laced.

Poland
White crested
Black Female

Plumage, Female: Except that the wing primaries are tipped, the colours and markings, including the crest, are buff ground and white lacing.

In both sexes: Beak dark blue or horn. Eyes, comb and face red. Ear-lobes blue-white. Legs and feet dark blue.

Plumage, Gold Variety, Male: Golden bay ground with black markings. Crest black at roots and tips, and as free as possible of whole white feathers. Muffling mottled or laced, not solid black. Hackle tipped. Back and saddle distinctly laced or spangled at the tips. Breast, thighs, shoulders and wings laced except primaries (of wings) which are tipped. Tail laced, the ends of the sickles well splashed.

Plumage, Female: Golden bay ground with black lacing, each feather distinctly marked and as free as possible from splashes.

In both sexes, beak, etc., as in the Chamois.

Plumage, Silver Variety, Male and Female: Plumage as in the Gold, substituting silver as the ground colour.

In both sexes, beak, etc., as in the Chamois.

Plumage, White Variety, Male and Female: Pure white.

In both sexes: Beak dark blue. Ear-lobes white. Other points as in the Chamois.

Plumage, White-crested Black Variety, Male and Female: Rich metallic black, except the crest which is snow white, with a black band at base of crest in front.

In both sexes, beak, etc., as in the White.

Plumage, White-crested Blue Variety, Male and Female: Solid dark blue (self coloured). Crest snow white with a blue band at base of crest in front.

In both sexes, beak, etc., as in the White.

Standard Weights: Male 6½ lb. Female 5 lb.

SCALE OF POINTS

White-crested Varieties

TYPE	5
HEAD (CREST 30; COMB AND WATTLES 15)	45
COLOUR	30
CONDITION	15
SIZE	5
	100

Other Varieties

TYPE....................................... 10
HEAD (CREST 30; MUFFLING 10)........ 40
COLOUR AND MARKINGS................. 30
CONDITION............................. 10
SIZE................................... 10

100

Serious Defects: Split or twisted crest. Comb, if any, other than horn type. Absence of muffling in Black, Chamois, Gold, Silver, and White varieties. Legs other than blue or slate. Other than four toes on each foot. Any deformity. Absence of black or blue in the front of the crest in White-crested Blacks and Blues.

REDCAP

ORIGIN: British
CLASSIFICATION: Light
EGG COLOUR: White

THE Redcap has always been closely associated with Derbyshire and is known as the Derbyshire Redcap. It was a sturdy breed carrying excellent breast meat, and a good egg producer. Farmers used the males freely for crossing to produce layers. As with so many promising utility breeds, the Redcap was bred and exhibited as if its immense comb was all that mattered, headpoints claiming 45 per cent of the 100 judging points.

GENERAL CHARACTERISTICS: MALE

Carriage: Graceful action and well balanced.

Type: Back broad, moderate in length falling slightly to tail and flat. Breast broad, full, and rounded. Wings moderate in length, neat and fitting closely to the body. Tail full and carried at an angle of about 60 degrees; broad, long and well-arched sickles.

Head: Of medium length and broad. Beak medium in length. Eyes full and prominent. Comb rose with straight leader, full of fine work or spikes, free from hollow in centre, set straight on the head and carried well off the eyes and beak; size about 3¼ by 2¾ inches. Face smooth and of fine texture. Ear-lobes of medium size. Wattles of medium length, well rounded and fine in texture.

Neck: Of moderate length, nicely arched, and with full hackle.

Legs and Feet: Legs straight and wide apart. Thighs short and well fleshed, shanks moderately long. Toes, four, well spread.

FEMALE

The general characteristics are similar to those of the male, allowing for the natural sexual differences, with the exception of the comb which is about half the size of the male's. The tail is large and full, and, like that of the male, carried at an angle of about 60 degrees.

COLOUR

Plumage, Male: Neck, hackle and saddle to harmonise. Each feather to have a red quill with beetle-green webbing, very finely fringed and tipped with black (the feathers appear to be fringed with red, but if placed on paper they are seen to have a black fringe and tip; the fringe is almost as fine as a hair). Back rich

Redcap Male

red, tipped with black. Wing bows rich red. Coverts rich red, each feather ending with a black spangle, forming a black bar across the wing. Primaries and secondaries black on one side red on the other side, heavily tipped with black. Breast and underparts black. Tail and hangers black.

Plumage, Female: Hackle as in the male, but nut-brown quill. Back and breast ground colour deep rich nut-brown, free from smuttiness, each feather ending with a half-moon black spangle. The markings on breast, back and wings to be as uniform as possible. Wing primaries and secondaries, as in the male, wing coverts evenly spangled. Tail black.

In both sexes: Beak horn. Eyes red. Comb, face, ear-lobes and wattles bright red. Legs and feet lead colour.

Standard Weights: Cock 6 to 6½ lb; cockerel 5½ to 6 lb. Hen 5 to 5½ lb; pullet 4½ to 5 lb.

SCALE OF POINTS

SIZE (STYLE AND SHAPE)	10
TAIL	5
HEAD POINTS (COMB 25, EYES 5, EAR-LOBES AND WATTLES 15)	45
LEGS AND FEET	5
COLOUR	25
CONDITION	10
	100

Serious Defects (for which birds should be passed): Comb over. White ear-lobes. Round backs. Squirrel or wry tails. Feathers on legs. Legs other colour than lead. Other than four toes. Crooked breast. Coarseness. Excessive fat.

RHODE ISLAND RED

ORIGIN: American
CLASSIFICATION: Heavy
EGG COLOUR: Light brown to brown

No breed made such a world progress in so short a time as this American breed. It was developed from Asiatic black-red fowls of the Shanghai, Malay, and Java types, bred on the farms of Rhode Island Province. Red Javas were known there in 1860, and the original Rhode Island Red had a rose comb, although birds with single, probably from Brown Leghorn crossings, and pea combs were also bred. The rose combs were first called American Reds and so standardised in 1905, a year after the single comb was accepted as the Rhode Island Red. However, in 1906 both combs became standardised as the Rhode Island Red. The formation of the British Rhode Island Red Club took place in 1909, and the breed has been one of the most popular in this country for all purposes. Being a gold, males of the breed are utilised extensively in gold-silver sex-linked matings.

GENERAL CHARACTERISTICS: MALE

Carriage: Alert, active, and well balanced.

Type: Body deep, broad and long; the keel bone long, straight, and extending well forward and back, giving the body an oblong look. Back broad, long and in the main nearly horizontal, this being modified by slightly rising curves at hackle and lesser tail coverts. Saddle feathers of medium length and abundant. Breast broad, deep and carried in a line nearly perpendicular with the base of the beak; at least it should not be carried farther back. Fluff moderately full, but with the feathers carried fairly close to the body; not a Cochin fluff. Wings of good size, well folded, and the flights carried horizontally. Tail of medium length, quite well spread, carried fairly well back, increasing the apparent length of the bird. Sickles of medium length, passing a little beyond the main tail feathers. Lesser sickles and tail coverts of medium length and fairly abundant.

Head: Of medium size, carried horizontally and slightly forward. Beak medium in length and slightly curved. Eyes full bright and prominent. Comb single or rose. The single of medium size, fine texture, set firmly in the head, perfectly straight and upright, with five even and well defined serrations, those in front and rear

146

Rhode Island Red
Male and Female

smaller than the centre ones, of considerable breadth where it is fixed to the head. The rose of medium size, low, set firmly on the head, the top oval in shape, and the surface covered with small points, terminating in a small spike at the rear. The comb to conform to the general curve of the head. Face smooth and of fine texture. Ear-lobes fairly well developed. Wattles medium and equal in length, moderately rounded and of fine texture.

Neck: Of medium length, carried slightly forward, and covered with abundant hackle, flowing over the shoulders but not too loosely feathered.

Legs and Feet: Legs well apart. Thighs large, of medium length and well covered with feathers. Shanks of medium length, well rounded and smooth. Toes, four, of medium length, straight, strong and well spread.

FEMALE

The general characteristics are similar to those of the male, allowing for the natural sexual differences. The tail, however, should not form an apparent angle with the back, nor must it be met by a high rising cushion. It should be a little shorter than medium and quite well spread. Neck hackle should be sufficient, but not too coarse in feather. In the mature hen the back would be described as broad, while in the pullet it would look somewhat narrower in proportion to the length of her body. The curve from the horizontal back to the hackle or tail should be moderate and gradual.

COLOUR

Plumage, Male: The neck red, harmonising with back and breast. Wing primaries, the lower web black and the upper red; secondaries, the lower web red and the upper black; flight coverts black; wing bows and coverts red. Tail, main feathers, including the sickles, black or greenish-black; coverts mainly black, but they may become russet or red as they approach the saddle.

The general surface of the plumage should be a rich brilliant red, except where black is specified. It should be free from shafting, mealy appearance or brassy effect. Absolute evenness of colour is desired.

The bird should be so brilliant in lustre as to have a glossed appearance. The undercolour and quill of the feather should be red or salmon. With the saddle parted, showing the undercolour at the base of the tail, the appearance should be red or salmon,

148

not whitish or smoky. Black or white in the undercolour of any section is undesirable.

Plumage, Female: Neck hackle red, the tips of the lower feathers having black ticking, but not heavy lacing. The tail should be black or greenish-black. In all sections of the wing the undercolour and quills of the feathers are as in the male. With the remainder of the plumage the surface should be a rich dark, even and lustrous red, but not as brilliant a lustre as in the male. It should be free from shafting or mealy appearance.

In both sexes, other things being equal, the specimen having the richest undercolour shall receive the award.

In both sexes: Beak red-horn or yellow. Eyes red. Face, comb, wattles and ear-lobes bright red. Legs and feet yellow or red-horn.

Standard Weights: Cock 8½ lb; cockerel 8 lb. Hen 6½ lb; pullet 5½ lb.

SCALE OF POINTS

SHAPE, SIZE, CARRIAGE AND SYMMETRY	30
COLOUR (GENERAL)	20
QUALITY AND TEXTURE (GENERAL)	15
HEAD AND COMB	10
EYE COLOUR	10
CONDITION	10
LEGS	5
	100

Serious Defects: Feather or down on shanks or feet, or unmistakable indications of a feather having been plucked from them. Badly lopped comb, side sprig or sprigs on the single comb. Other than four toes. Entire absence of main tail feathers. Two absolutely white (so-called wall or fish) eyes. Squirrel and wry tail. A feather entirely white that shows in the outer plumage. An ear-lobe showing more than one-half of the surface permanently white. (This does not mean the pale ear-lobe, but the enamelled white.) Diseased specimens, crooked backs, deformed beaks, shanks and feet other than yellow or red-horn colour. A pendulous crop shall be cut hard. Coarseness. Toes not straight and well spread. Super-fineness. Under all disqualifying clauses, the specimen shall have the benefit of the doubt.

Robustness is of vital importance.

SCOTS DUMPY

ORIGIN: British
CLASSIFICATION· Light
EGG COLOUR: White

THIS breed has been bred in Scotland for more than a hundred years, and the birds known also as Bakies, Crawlers and Creepers. Fowls having identical dumpy characters have been described as early as 1678. The breed is considered an ideal sitter and mother.

GENERAL CHARACTERISTICS: MALE

Carriage: Heavy, with a waddling gait, the extreme shortness of its legs giving the bird the appearance of " swimming on dry land ". Shortness of leg alone should not constitute the breed's claim to notice. The large, low, heavy body, and other points of excellence must be possessed also.

Type: Body square. Back broad and flat. Breast deep. Wings of medium size and neatly carried. Tail full and flowing, the sickles well arched.

Head: Fine. Beak strong and well curved. Eyes large and clear. Comb single or rose, the former preferred. The single of medium size, upright and straight, free from side sprigs, and the back following the line of the skull, evenly serrated on top. The rose also of medium size, straight and firmly set, full of fine work or spikes, level on top, and narrowing behind to a distinct peak following the line of the skull and not sticking up. Face smooth. Ear-lobes small and close to the neck. Wattles of medium size.

Neck: Of fair length, in keeping with the size of the body, and covered with flowing hackle.

Legs and Feet: Legs very short, the shanks not exceeding $1\frac{1}{2}$ inches. Toes, four, well spread.

FEMALE

The general characteristics are similar to those of the male, allowing for the natural sexual differences.

150

Scots Grey
Male

Scots Dumpy
Male

COLOUR

Plumage, Male and Female: There is no fixed plumage colour, but the varieties chiefly exhibited are Black, Cuckoo, Dark, and Silver-Grey, the last three being similar to those varieties of the Dorking.

In both sexes: Eyes red. Comb, face, wattles and ear-lobes bright red. Beak, legs and feet white, except in the Black variety where they should be black or slate, and in the Cuckoo mottled.

Standard Weights: Male 7 lb. Female 6 lb.

SCALE OF POINTS

TYPE	40
SIZE	20
HEAD	15
CONDITION	15
COLOUR	10
	100

Serious Defects: White ear-lobes. Yellow or feathered shanks or feet. Long legs. Any deformity.

SCOTS GREY

ORIGIN: British
CLASSIFICATION: Light
EGG COLOUR: White

A LIGHT non-sitting breed originated in Scotland, it has not been bred extensively outside that country where, even if it is less popular today, it will doubtless be maintained by keen breeders. It has been bred there for over two hundred years.

GENERAL CHARACTERISTICS: MALE

Carriage: Erect, active and bold.

Type: Body compact, full of substance and fairly long. Back broad and flat. Breast deep, full and carried upwards. Wings moderately long and well tucked, the bow and tip covered by the neck and saddle hackles. Tail fairly long and well up (but not squirrel fashion) with full sickles.

Head: Long and fine. Beak strong and well curved. Eyes large and bright. Comb single, upright, of medium size, with well defined serrations, the back following the line of the skull. Face of fine texture. Ear-lobes of medium size. Wattles of medium length with a well-rounded lower edge.

Neck: Finely tapered and with profuse hackle flowing on the back and shoulders.

Legs and Feet: Legs long and strong. Thighs wide apart but not quite as prominent as those of Game fowl. Shanks free from feathers. Toes, four, straight and spreading, stout and strong.

Handling: Firm and somewhat similar to the Game fowl.

FEMALE

With the exception of the comb, either erect or falling slightly over, the general characteristics are similar to those of the male, allowing for the natural sexual differences.

COLOUR

Plumage, Male: Cuckoo-feathered. Ground colour of body, thighs and wing feathers blue-white, but that of the neck hackle, saddle and tail may vary from blue-white to light grey. The barring is black with a metallic lustre, that of the body, thighs and wing feathers straight across, but that of the neck hackle, saddle, and tail slightly angled or V-shaped. The alternating bands of black are of equal width and proportioned to the size of the feather. The bird should " read " throughout, i.e., the shade should be the same from head to tail. The plumage should be free from red, black, white, or yellow feathers, and the hackle, saddle, and tail should be distinctly and evenly barred, while the markings all over should be rather small, even, and sharply defined.

Plumage, Female: Similar to that of the male, except that the markings are not as small, and produce an appearance somewhat resembling a shepherd's tartan.

In both sexes: Beak white or white streaked with black. Eyes amber. Comb, face, wattles and ear-lobes bright red. Legs and feet white or white mottled with black, but not sooty.

Standard Weights: Male 7 lb. Female 5 lb.

SCALE OF POINTS

COLOUR AND MARKINGS (HACKLE 10, BACK
    10, TAIL 10, WINGS AND ACROSS SHOUL-
    DERS 10, BREAST AND THIGHS 10)..... 50
SIZE.. 15
TYPE....................................... 10
HEAD....................................... 10
CONDITION.................................. 10
LEGS AND FEET.............................. 5
                                                      —
                                                    100

Serious Defects: Any bodily deformity. Any characteristic of any other breed not applicable to the Scots Grey.

SILKIE

ORIGIN: Asiatic
CLASSIFICATION: Light
EGG COLOUR: Tinted to cream

SILKIE fowls have been mentioned by authorities for several hundred years, although some think they originated in India, while others favour China and Japan. It seems likely that such oddities came from Japan. Despite such small weights of 2 to 3 lb the Silkie is not regarded as a bantam in this country but as a light breed, and as such it must be exhibited. Its persistent broodiness is a breed characteristic, and either pure or crossed the breed provides reliable broodies for the eggs of large fowl or bantams.

GENERAL CHARACTERISTICS: MALE

Carriage: Stylish, compact and lively.

Type: Body broad and stout looking. Back short, saddle silky and rising to the tail, stern broad and abundantly covered with fine fluff, saddle hackles soft, abundant, and flowing. Breast broad and full; shoulders stout, square, and fairly covered with neck hackle. Wings soft and fluffy at the shoulders, the ends of the flights ragged and " Osprey plumaged " (i.e., some strands of the flight hanging loosely downward). Tail short and very ragged at the end of the harder feathers of the tail proper. It should not be flowing, but a short round curve.

Silkie
White Male,
Black Female

Head: Short and neat, with good crest, soft and full, as upright as the comb will permit, and having half a dozen to a dozen soft, silky feathers streaming gracefully backwards from lower and back part of crest, to a length of about an inch and a half. The crest proper should not show any hardness of feathers. Beak short and stout at base. Eyes brilliant and not too prominent. Comb almost circular in shape, preferably broader than long, with a number of small prominences over it; preferably having a slight indentation or furrow transversely across the middle. Face smooth. Ear-lobes more oval than round. Wattles concave, nearly semi-circular, not long or pendent.

Neck: Short or medium length, broad and full at base with the hackle abundant and flowing.

Legs and Feet: Free from scaliness. Thighs wide apart and legs short. No hard feathers on the hocks but a profusion of soft silky plumage on them is admissible. The feathers on the legs should be moderate in quantity. Toes, five, the fourth and fifth diverging from one another. The middle and outer toes feathered, but these feathers should not be too hard. Thighs covered with abundant fluff.

Plumage: Very silky and fluffy with a profusion of hair-like feathers.

FEMALE

Saddle broad and well cushioned with the silkiest of plumage which should nearly smother the small tail, the ragged ends alone protruding, and inclined to be " Cochiny " in appearance. The legs are particularly short in the female, in which the under fluff and thigh fluff should almost meet the ground. The head crest is short and neat, like a powder puff, with no hard feathers, nor should the eye be hidden by the crest, which should stand up and out, not split by the comb. Ear-lobes small and roundish. Wattles either absent or small and oval in shape. Other general characteristics are similar to those of the male, allowing for the natural sexual differences. Comb small.

COLOUR

Plumage, Black Variety, Male and Female: Black all over with a green sheen in the males; colour in hackle is permissible.

Plumage, Blue Variety, Male and Female: An even shade of blue from head to tail; a self-colour, and neither laced nor barred.

Plumage, Gold Variety, Male and Female: An even shade of golden buff, avoiding pale lemon colour on the one hand and brownish orange on the other. Clear colour throughout to be preferred but some darker feathers permissible in tails of both sexes.

Plumage, White Variety, Male and Female: Snow white.

In both sexes: Beak slaty-blue. Eyes black. Comb, face and wattles mulberry. Ear-lobes turquoise-blue or mulberry, the former preferred. Legs and feet lead. Nails blue-white. Skin mulberry.

Standard Weights: Male 3 lb. Female 2 lb. (There is no objection to size in either sex if the other points are good.)

SCALE OF POINTS

TYPE	20
HEAD	30
LEGS	10
COLOUR	10
PLUMAGE	30
	100

Serious Defects: Hard feathers. Vulture hocks. Green beak or green tip to beak. Horns protruding from comb. Ruddy comb or face. Eye other than black. Incorrect colour in plumage or skin. Plumage not silky. Want of crest; " Polish " or " split " crest; the crest should not hang over the eyes. Green soles to feet.

Disqualifications: Single comb. Green legs. Four toes. Featherless legs and feet.

SPANISH

ORIGIN: Mediterranean
CLASSIFICATION: Light
EGG COLOUR: White

THE White-faced Black Spanish is one of our oldest breeds, and was widely kept and admired long before the advent of poultry shows in the latter half of the nineteenth century. Of striking appearance, with its extensive white face, surrounding eyes and ears and extending lower than the wattles, the Spanish was also a good layer of large white eggs. The emergence of the red-faced Minorca, pushed the Spanish into the background, but they still appear occasionally at shows.

GENERAL CHARACTERISTICS: MALE

Carriage: Upright, with proud action.

Head: Skull long, broad and deep. Beak long and stout. Eyes full and wide open. Comb single, somewhat small, erect and straight, firm at the base, rather thin at the edge, fitting closely on the neck at the back, of very smooth texture, and free from wrinkles, rising well over the eyes but not so as to interfere with the sight, and joining the ear-lobes and wattles. Ear-lobes deep and broad, well rounded at the bottom, extending well below the wattles, meeting in front and going well back on each side of the neck, of fine texture and free from folds or creases. Wattles very long, thin, and pendulous.

Neck: Long and fine, with abundant hackle flowing well over the shoulders.

Body: Rather long, fairly broad in front, and tapering to the rear. Breast full at the neck and gradually decreasing towards the thighs. Back slanting downwards to the tail, short wings carried closely. Full tail, not carried too high, and with the sickles large and well curved.
Legs: Rather long and slim. Shanks free of feathers. Toes (four) slender and straight.
Plumage: Short and close.

FEMALE

With the exception of the comb (which falls gracefully over either side of the face) the general characteristics are similar to those of the male, allowing for the natural sexual differences.

COLOUR

Beak: Dark horn.   Eyes black.   Comb and wattles bright red. Face and ear-lobes white.   Legs and feet pale slate.

Plumage: Black with a beetle-green sheen, and free of purple bars.

Standard Weights: Male 7 lb.   Female 6 lb.

SCALE OF POINTS

FACE AND LOBES	35
COMB AND WATTLES	15
TYPE..	15
SIZE	15
COLOUR	10
CONDITION	10
	100

Serious Defects: Blue, pink or red in face or lobes.   Coarse " cauliflower " face or lobes.   Cock's comb not erect, side sprigs on comb.   Lobes pointed at the bottom.   Black or dark legs or feet. Any deformity.

SULTAN

ORIGIN: Turkish
CLASSIFICATION: Light
EGG COLOUR: White

SULTAN fowls were imported in 1854 from Constantinople. Though they never became widespread, Sultans have remained in existence and are occasionally seen at shows.   In crest, beard and plumage they are similar to White Polands.   However, they differ in some important respects: vulture hocks, five toes and short back. The Sultan is a breed of considerable character, tame, yet sprightly.

GENERAL CHARACTERISTICS: MALE

General Shape and Carriage: Deep, but neat and compact, and very sprightly.

Head and Neck: Head medium size. Beak short and curved. Eye bright. Comb very small, consisting of two spikes only, almost hidden by crest. Face covered with thick muffling. Nostrils horny and large, rising above the curved line of the beak. Crest large, globular, and compact. Ear-lobes small and round. Beard very full, joining with the whiskers. Wattles very small, to be hardly perceptible. Neck moderately short, slightly arched and carried well back.

Body: Rather long and very deep. Breast deep and prominent. Back short and straight. Wings large, long, and carried low.

Tail: Long, broad, and carried open. Sickles very long and fine. Hangers numerous, long and fine. Coverts abundant and lengthy.

Legs and feet: Thighs short, furnished with heavy vulture hocks to cover the joints. Shanks short, and well covered inside and out with feathers. Toes five in number and of moderate length, completely covered with feather.

Size and Weight: Medium to about 6 lb.

Plumage: Long, very abundant and fairly soft.

FEMALE

With the exception of the comb, which is smaller and barely visible, the general characteristics are similar to those of the male, allowing for the natural sexual differences. Weight $4\frac{1}{2}$ lb.

COLOUR

In both sexes: Beak pale blue or white. Eye red. Comb bright red. Face red. Ear-lobes and wattles bright red. Shanks and toes pale blue. Plumage snow white throughout.

SCALE OF POINTS

HEAD AND CREST........................... 15
BEARD AND MUFFLING 15
COMB 5
TYPE AND SYMMETRY................... 12
COLOUR 15
LEG AND FOOT-FEATHERING 15
SIZE 8
CONDITION 15
                                             ———
                                             100

Serious Defects: Any deformity, coloured plumage, toes other than five in number.

SUMATRA GAME

ORIGIN: Asiatic
CLASSIFICATION: Light
EGG COLOUR: White

SUMATRA GAME were admitted to the American Standard in 1883, under the name of Black Sumatra. They reached Britain later and have been bred here in small numbers ever since. Its long flowing tail, carried horizontally, and its pheasant-like carriage are two of the main characteristics. Sumatra Game are prolific layers, and excellent sitters. (For photograph see page 205.)

GENERAL CHARACTERISTICS: MALE

Carriage: Straight and upright in front, pheasant-like, giving a proud and stately appearance.

Head: Skull small, rather short, and somewhat rounded. Beak strong, of medium length, slightly curved. Eyes large and very bright, with a quick and fearless expression. Comb pea, low in front, fitting closely, the smaller the better. Face smooth and of fine texture. Ear-lobes and wattles as small as possible and fitting very closely.

Neck: Rather long, and covered with very long and flowing hackle.

Body: Rather long, very firm and muscular, broad, full, and rounded breast. Back of medium length, broad at shoulders, very slightly tapering to tail. Saddle hackle very long and flowing. Stern narrower than shoulders, but firm and compact. Strong, long, and large wings, carried with fronts lightly raised, the feathers folded very closely together, not carried drooping or over the back. Long drooping tail with a large quantity of sickles and coverts, which should rise slightly above the stern and then fall streaming behind, nearly to the ground. Sickle and covert feathers not too broad.

Legs: Of strictly medium length, thick and strong. Thighs muscular, set well apart. Shanks straight and strong, set well apart, with smooth, even scales, not flat or thin. (*Note:* There is no objection to two or more spurs on each leg, it being a peculiarity of the breed for this to occur.) Feet broad and flat. Toes (four) long, straight, spread well apart, with strong nails, the back toe standing well backward and flat on the ground.

Plumage: Very full and flowing, but not soft or fluffy.

FEMALE

The general characteristics are similar to those of the cock, allowing for the natural sexual differences.

COLOUR

Beak: Dark olive or black (olive preferred). Eyes very dark red, dark brown, or black (dark red preferred).

Face, comb, ear-lobes and wattles: Black or " gipsy " faced or very dark red (gipsy face preferred).

Legs and feet: Dark olive or black (olive preferred).

Plumage: Very rich beetle-green (green-black) with as much sheen as possible.

Standard Weights: Male 5 lb.   Female 4 lb.

SCALE OF POINTS

TYPE	20
HEAD (BEAK 5, EYES 5, OTHER POINTS 10)	20
COLOUR	15
FEATHER, QUANTITY OF	15
CONDITION	15
LEGS AND FEET	10
NECK	5
	100

Serious Defects: Single or rose comb, dubbing. Other than four toes. Any deformity.

SUSSEX

ORIGIN: British
CLASSIFICATION: Heavy
EGG COLOUR: Tinted

THIS is a very old breed, for although we do not find it included in the first Book of Standards of 1865, yet at the first poultry show of 1845 the classification included Old Sussex or Kent Fowls, Surrey Fowls and Dorkings. The oldest variety of the Sussex is the Speckled. Brahma, Cochin, and Silver-grey Dorking were used in the make-up of the Light. The earlier Reds had black breasts, until the Red and Brown became separate varieties. Old English Game has figured in the make-up of some strains of Browns. Buffs appeared about 1920, clearly obtained by sex-linkage within the breed. Whites came a few years later, as sports from Lights. Silvers are the latest variety. The Light is the most widely kept in this country today, among Standard as well as commercial breeders. It is one of our most popular breeds for producing table birds. At the time when sex-linkage held considerable popularity the Light Sussex was one of the most popular breeds of the day, the females being in considerable demand for mating to gold males. At an even earlier stage the Sussex breed formed the mainstay of the table poultry market in and around the Heathfield area. The Sussex Breed Club was formed as far back as in 1903 and is now one of the oldest breed clubs in Britain.

GENERAL CHARACTERISTICS: MALE

Carriage: Graceful, showing length of back, vigorous and well balanced.

Type: Back broad and flat. Breast broad and square, carried well forward, with long and straight, deep, breast bone. Shoulders wide. Wings carried close to the body. Skin clear and of fine texture. Tail moderate size, carried at an angle of 45 degrees.

Head: Of medium size and fine quality. Beak short and curved. Eyes prominent, full and bright. Comb single, of medium size, evenly serrated and erect, and fitting close to the head. Face smooth and of good texture. Ear-lobes and wattles of medium size and fine texture.

163

Light Sussex
Male and Female

Sussex
Speckled Male
and Female

Neck: Gracefully curved with fairly full hackle.

Legs and Feet: Thighs short and stout. Shanks short and strong, and rather wide apart, free from feather, with close fitting scales. Toes, four, straight and well spread.

Plumage: Close and free from any unnecessary fluff.

FEMALE

The general characteristics similar to those of the male, allowing for the usual sexual differences.

COLOUR

Plumage, Brown Variety, Male: Head and neck hackles rich dark mahogany striped with black. Saddle hackle same as neck hackle. Back and wing bow rich dark mahogany. Wing coverts forming the bar, blue-black; secondaries and flights black, edged with brown. Breast, tail and thighs black.

Plumage, Female: Head and neck hackles brown striped with black. Back and wings dark brown, finely peppered with black. Breast and underbody clear pale wheaten brown. Flights black, edged with brown. Tail black.

Plumage, Buff Variety, Male and Female: Body rich even golden buff. Head and neck hackles buff, sharply striped with green-black. Wings buff, with black in the flights. Tail and coverts greeny-black. Dark in undercolour, not penalised at present, but buff is desirable.

Plumage, Light Variety, Male and Female: Head and neck hackles white, striped with black, the black centre of each feather to be entirely surrounded by a white margin. Wings white, with black in flights. Tail and coverts black. Remainder pure white throughout.

Plumage, Red Variety, Male and Female: Head and neck hackles rich dark red, striped with black. Body and wing bow rich dark red, one uniform shade throughout free from pepperiness. Wings rich dark red with black in the flights. Tail black, coverts rich dark red. Undercolour slate.

Plumage, Speckled Variety, Male: Head and neck hackles rich dark mahogany, striped with black and tipped with white. Wing bow speckled, primaries white, brown and black. Saddle hackle similar to neck hackle. Tail, main feathers black and white, sickles black with white tips. Remainder rich dark mahogany, each feather tipped with a small white spot, a narrow glossy black bar dividing the white from the remainder of the feather. Undercolour slate and red with a minimum of white.

166

Sussex
Silver Male
Brown Female

Plumage, Female: Head, neck and body ground colour rich dark mahogany, each feather tipped with a small white spot, a narrow glossy black bar dividing the white from the remainder of feathers, the mahogany part of feather free from pepperiness, neither of the colours to run into each other, and to show the three colours distinctly; undercolour as for male. Tail black and brown with white tip. Flights black, brown and white.

Plumage, Silver Variety, Male: Head, neck and saddle hackles white striped with black, the black centre of each feather to be entirely surrounded by a white margin. Wing bow and back silvery white; coverts forming bar black; flights and secondaries black tinged with grey. Breast black with white shafts, and silver lacing round feathers. Thighs dark grey showing faint lacing. Tail black. Undercolour grey-black shading to white at skin.

Plumage, Female: Head and neck hackles as in the male. Back and wing bow greyish black, each feather showing white shaft with fine silver lacing surrounding it; flights and secondaries greyish black. Tail black. Breast and thighs lighter shade of greyish black with white shafts and silver lacing to correspond with the top colour. Undercolour as in the male.

Plumage, White Variety, Male and Female: Pure white throughout and to the skin.

In both sexes: Beak white or horn generally, dark or horn with the Brown and dark shading to white with the Silver. Eyes: Brown—brown or red; Buff, Red and Speckled—red; Light, Silver and White—orange. Face, comb, ear-lobes and wattles red. Shanks and feet white. Flesh and skin white.

Standard Weights: Male 9 lb (min.). Female 7 lb (min.).

SCALE OF POINTS

TYPE AND FLATNESS OF BACK	25
SIZE	20
LEGS AND FEET	15
COLOUR	20
CONDITION	10
HEAD AND COMB	10
	100

Serious Defects: (for which birds should be passed.) Other than four toes. Wry tail or any other deformity. Feather on shanks and toes. Rose comb.

WELSUMMER

ORIGIN: Dutch
CLASSIFICATION: Light
EGG COLOUR: Brown to deep brown

NAMED after the village of Welsum, this Dutch breed has in its make-up such breeds as the Partridge Cochin, Partridge Wyandotte and Partridge Leghorn, and still later the Barnevelder and the Rhode Island Red. In 1928, stock was imported into this country from Holland, in particular for its large brown egg, which remains its special feature, some products being mottled with brown spots. It has distinctive markings and colour, and comes into the light-breed category, although it has good body-size. It enters the medium class in the country of its origin. Judges and breeders work to a Standard that values indications of productiveness, so that laying merits can be combined with beauty.

GENERAL CHARACTERISTICS: MALE

Carriage: Upright, alert and active.

Type: Body well built on good constitutional lines. Back broad and long. Breast full, well rounded and broad. Wings moderately long, carried closely to the sides. Tail fairly large and full, carried high, but not squirrel. Abdomen long, deep and wide.

Head: Symmetrical, well balanced, of fine quality without coarseness, excesses or exaggeration. Skull refined, especially at back. Beak strong, short and deep. Eyes keen in expression, bold, full, highly placed in skull and standing out prominently when viewed from front or back; pupils large and free from defective shape. Comb single, of medium size, firm, upright, free from any twists or excess around nostrils, clear of nostrils, and of fine, silky texture, five to seven broad and even serrations, the back following closely but not touching the line of the skull and neck. Face smooth, open and of silky texture, free from wrinkles or surfeit of flesh and without overhanging eyebrows. Ear-lobes small and almond shaped. Wattles of medium size, fine and silky texture and close together.

Neck: Fairly long, slender at top but finishing with abundant hackle.

Legs and Feet: Thighs to show clear of body without loss of breast. Shanks of medium length, medium bone and well set apart, free from feathers and with soft, pliable sinews, free from coarseness. Toes, four, long, straight and well spread out, back toe to follow in straight line, free from feathers between toes.

Plumage: Tight, silky and waxy, free from excess or coarseness, silky at abdomen and free from bagginess at thighs.

Handling: Compact, firm and neat bone throughout.

FEMALE

The general characteristics are similar to those of the male, allowing for the natural sexual differences. Handling: Pelvic bones fine and pliable; abdomen pliable; flesh and skin of fine texture and free from coarseness; plumage sleek; abdomen capacious, but well supported by long breast-bone and not drooping; general handling of a fit, keen and active layer.

COLOUR

Plumage, Male: Head and neck rich golden brown. Hackles rich golden brown as uniform as possible, free from black striping, yet underparts (out of sight) may show a little striping at present. Back, shoulder coverts and wing bow bright red-brown. Wing coverts black with green sheen forming a broad bar across (a little brown peppering at present permissible); primaries (out of sight when wing is closed), inner web black, outer web brown; secondaries, outer web brown, inner web black with brown peppering. Tail (main) black with a beetle-green sheen; coverts, upper black, lower black edged with brown. Breast black with red mottling. Abdominal and thigh fluff black and red mottled.

Plumage, Female: Head golden brown. Hackle golden brown or copper, the lower feathers with black striping and golden shaft. Breast rich chestnut red going well down to the lower parts. Back and wing bow reddish-brown, each feather stippled or peppered with black specks (i.e., partridge marking), shaft of feather showing lighter and very distinct. Wing, bar chestnut brown; primaries, inner web black, outer brown; secondaries, outer web brown, coarsely stippled with black; inner web black, slightly peppered with brown. Abdomen and thighs brown with grey shading. Tail black, outer feathers pencilled with brown.

In both sexes: Beak yellow or horn. Eyes red. Comb, face, ear-lobes and wattles bright red. Legs and feet yellow. Under-colour dark slate grey.

170

Welsummer
Male and Female

Plumage, Silver Duckwing Variety, Male: Head, neck and hackles, white. Breast, black with white mottling. Back shoulder coverts and wing bow, white. Wing primaries, flight feathers (out of sight as wing is closed), inner web black, outer web white; secondaries, outer web white, inner web black, with white peppering, coverts black with green sheen forming a broad bar across primaries. Tail, main black with beetle-green sheen; coverts, upper black, lower black, edged with white. Abdominal and thigh fluff, black with white mottling.

Plumage, Female: Head, and skull, silvery white. Hackle, silvery white and lower feathers with black striping, and white shaft. Breast, salmon red or robin red. Back and wing bow, silvery grey, each feather stippled or peppered with black specks (i.e., partridge marking), shaft of feather showing light and very distinct. Wing bar silvery grey; primaries, inner web black, outer web white; secondaries, outer web white, coarsely stippled with black, inner web, black slightly peppered with white. Abdomen and thighs, silvery grey. Tail black, outer feathers pencilled with white.

Standard Weights: Cock 7 lb; cockerel 6 lb. Hen 6 lb; pullet $4\frac{1}{2}$ to 5 lb.

SCALE OF POINTS

GENERAL TYPE	20
HANDLING, SIZE, AND INDICATIONS OF PRODUCTIVENESS	30
HEAD	10
LEGS AND FEET	10
COLOUR	20
CONDITION	10
	100

Serious Defects: Comb other than single or with side sprigs. White in lobe. Feather on legs, hocks or between toes. Other than four toes. Striping in neck hackle or saddle of male. Absolutely black or whole red breast in the male. Salmon breast in the female. Legs other than yellow. Badly crooked or duck toes. Any body deformity. Coarseness, beefiness and anything which interferes with the productiveness and general utility of the breed.

172

WYANDOTTE

Origin: American
Classification: Heavy
Egg Colour: Tinted

THE first variety of the Wyandotte family was the Silver Laced, originated in America, where it was standardised in 1883. The variety was introduced into England at the time, and our breeders immediately perfected the lacings and open ground colouring.

Partridge Cochin and Gold Spangled Hamburgh males were crossed with the Silver females, to produce the Gold Laced variety. The White Wyandotte came as a sport from the Silver Laced; the Buff followed by crossing Buff Cochin with the Silver Laced. In 1896 the Partridge variety was introduced from America, the result of blending Partridge Cochin and Indian Game blood with that of the Gold Laced, the variety being perfected for markings in England. It was once called the Golden Pencilled, and the Silver Pencilled soon followed from Partridge Wyandotte and Dark Brahma crossings.

Columbians were the result of crossing the White Wyandotte with the Barred Rock, and it was the crossing of the Gold Laced and the White varieties which produced the Buff Laced and the Blue Laced, first seen here in 1897. Blacks, Blues and Barred have been made in different ways in this country. The latest variety to be introduced is the Red, created in Lancashire, from the Gold Laced variety, with selective matings with White Wyandotte, Barnevelder and Rhode Island Red.

It is clear that while the family of the Wyandotte is large, every variety is a made one from various blendings of breeds.

GENERAL CHARACTERISTICS: MALE

Carriage: Graceful, well balanced, alert and active, but docile.

Type: Body short and deep with well rounded sides. Back broad and short with full and broad saddle rising with a concave sweep to the tail. Breast full, broad and round with a straight keel bone. Wings of medium size, nicely folded to the side. Tail medium size but full and spread at the base, the main feathers carried rather upright, the sickles of medium length.

Head: Short and broad. Beak stout and well curved. Eyes intelligent and prominent. Comb rose, firmly and evenly set on head, medium in height and width, low, and square at front, gradually tapering towards the back and terminating in a well-defined spike (or leader) which should follow the curve of the neck without any upward tendency. The top should be oval and covered with small and rounded points; the side outline being convex to conform to the shape of the skull. Face smooth and fine in texture. Ear-lobes oblong, wattles medium length, fine in texture.

Neck: Of medium length and well arched with full hackle.

Legs and Feet: Thighs of medium length, well covered with soft feathers; the fluff fairly close and silky. Shanks medium in length, strong, well rounded, good quality, and free of feather or fluff. Toes, four, straight and well spread.

Plumage: Fairly close and silky, not too abundant or fluffy.

FEMALE

The general characteristics are similar to those of the male, allowing for the natural sexual differences.

COLOUR

Plumage, Barred Variety, Male and Female: Similar to that of the Barred Plymouth Rock.

Plumage, Black Variety, Male and Female: Black with beetle-green sheen, undercolour as dark (black) as possible.

Plumage, Blue Variety, Male and Female: One even shade of blue, light to dark, but medium preferred; a clear solid blue, free from mealiness, " pepper ", sandiness, or bronze, and quite clear of lacing; a " self colour " in fact.

Plumage, Blue Laced Variety, Male: Bay (red-brown or chestnut) with blue markings. Hackles distinctly striped with blue down the centre of each feather and free from black tips or black around the edging. Back and shoulders free from black or smutty blue. Wing bar laced blue, well defined. Breast regularly and distinctly laced from the throat to the back of the thighs and free from double, outer, black or smutty marking. Fluff blue, powdered with gold. Tail solid blue, free from black or white.

Plumage, Female: Hackle and tail as in the male. Remainder as on the male's breast, the lacing extending to back of thighs into the fluff.

174

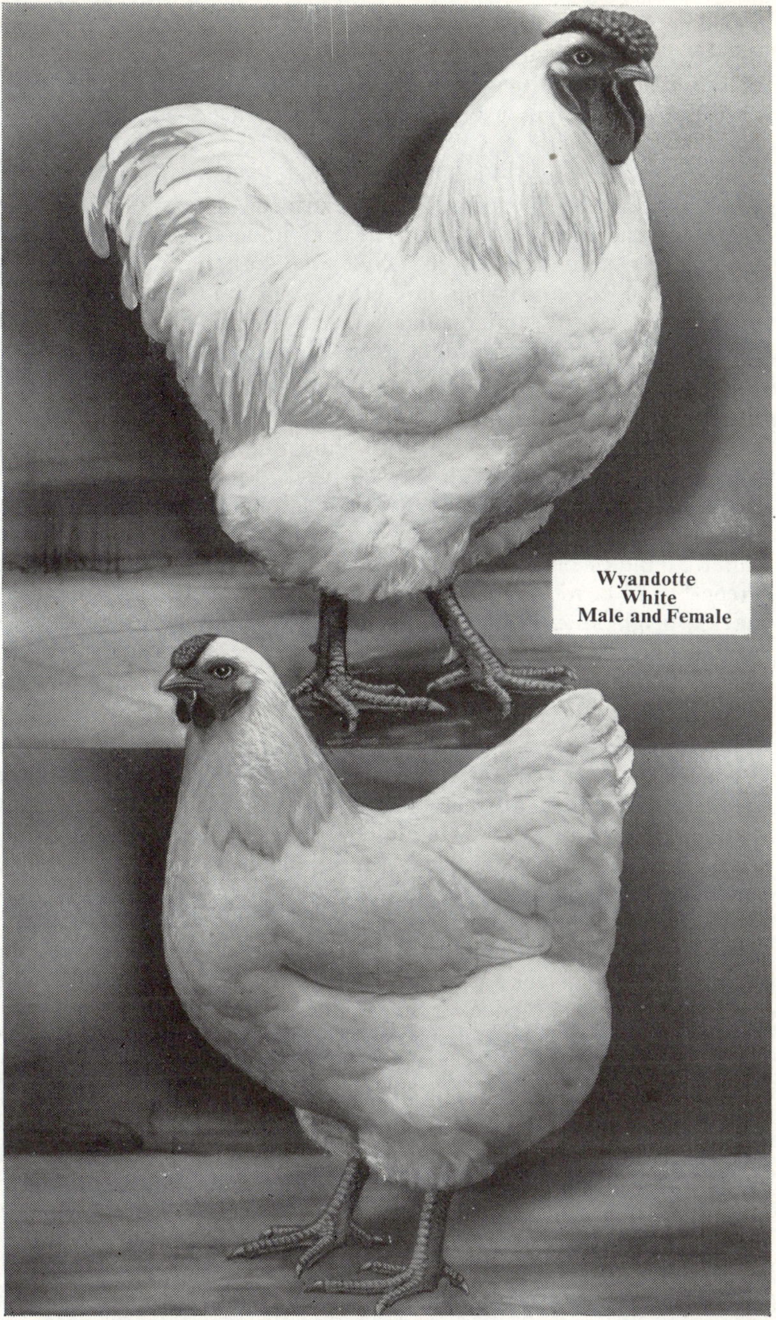

Wyandotte
White
Male and Female

Plumage, Buff Variety, Male and Female: Clear, sound buff throughout to skin, allowing greater lustre on the hackles and wing bow of the male. With these exceptions the colour should be perfectly uniform, but washiness or a red tinge, mealiness or " pepper " to be avoided.

Plumage, Buff Laced Variety, Male and Female: Rich buff with white markings. Hackles distinctly striped with white down the centre of each feather. Tail, fluff and undercolour white; wing secondaries, white inner web, outer web buff laced. The male's back, shoulders and wing bow rich solid buff. Remainder (in both sexes) clearly and regularly laced with white, the lacing of the female's cushion sometimes continuing into the tail coverts.

Plumage, Columbian Variety, Male and Female: Pearl-white with black markings; primaries (wing) black or black edged with white, secondaries black inner web and white outer. The male's neck hackle broadly striped with intense black down the centre of each feather, such stripe entirely surrounded by a clearly defined white margin and finishing with a decided white point (free from black outer edging or black tips); saddle hackle white, and the tail glossy green-black, coverts either laced or not with white. The female's hackle bright intense black, each feather entirely surrounded by a well defined white margin, and tail feathers black, except the top pair which may or may not be laced with white. Remainder (in both sexes) white, entirely free of ticking, with slate, blue-white or white undercolour.

Plumage, Gold Laced Variety, Male: Head, a rich golden bay, the neck hackle with a distinct black stripe down the centre of each feather, free from ticks, black outer edging or black tips. Saddle hackles to match neck. Back, a rich golden bay, free from black or maroon. Shoulder tip bay laced with black. Wing, bow bay, coverts evenly laced, forming at least two well defined bars; secondaries black on inner and wide bay stripe on outer web, the edge laced with black; primaries black on inner web and broadly laced bay on outer edge. Breast and underparts, the web bay, with well defined jet black lacing, free from double or bay outer lacing, the lacing regular from throat to back of thighs and showing a green lustre. Undercolour dark slate. Tail, the true tail feathers, sickle and coverts, black showing a green lustre. Thighs and fluff, black or dark slate, with clear lacing round hocks and outer side of thighs. Brightness and uniformity of colour to be considered of more value than any particular shade. Gold shaft permissible.

Plumage, Female: Head and neck hackle as in the male. Breast

Wyandotte
Columbian
Male and Female

and back, undercolour dark slate, web bay with regular well defined jet-black lacings, free from double or outer lacing, and showing a green lustre. Wings as the broad portion of the back; secondaries and primaries as in the male. Tail black, showing a green lustre, the coverts black with a bay centre to each feather. Thighs and fluff black or dark slate. Brightness and a quality of ground colour and regularity of lacing throughout are of first importance. Gold shaft permissible.

Plumage, Silver Laced Variety, Male and Female: Except that the ground is silver-white (free from yellow or straw tinge), instead of rich golden bay, the Silver Laced is similar to the Gold Laced. In the female, regularity of lacing and quality of colour must count above any particular breadth of lacing. Silver shaft permissible.

Plumage, Partridge Variety, Male: Head dark orange. Hackles bright orange-yellow, shading to bright lemon-yellow, free from washiness, each feather having a clearly defined glossy-black stripe down the middle, not running out at the tip, and free from light shaft. Back and shoulders bright red of a scarlet shade, free from maroon or purple tint. Wing bar solid, glossy black; primaries solid black, free from white; secondaries rich bay outer web and black inner and end of feather, the rich bay alone showing when the wing is closed. Undercolour black or dark grey, free from white. Breast and fluff metallic black, free from red or grey ticking. Tail (including sickles and tail coverts) metallic black, free from white at roots.

Plumage, Female: Head and hackle rich golden yellow, the larger feathers finely and clearly pencilled. Breast, back, cushion and wings soft light partridge-brown, quite even and free from red or yellow tinge, each feather plentifully and distinctly pencilled with black, the pencilling to follow the form of the feather, and to be even and uniform throughout. Fine, sharply defined pencilling with three or more distinct lines of black is preferred to coarse, broad marking, especially in hens, in which the pencilling is generally better defined than in pullets. Pencilling that runs into the brown, peppery markings, and uneven, broken or barred pencilling, constitute defects. Light shafts to feathers on the breast must be penalised. Fluff brown (same shade as body), as clearly pencilled as possible. Primaries (wing) black, secondaries brown (same shade as body), pencilled with black on outer web, black on inner web, showing pencilling when the wing is closed. Tail black, with or without brown markings, with clearly pencilled feathers up to the point of the tail.

Wyandotte
Partridge Male,
Black Female

Plumage, Silver Pencilled Variety, Male and Female: Except that the ground is silver-white in the male and steel-grey in the female, instead of red, brown, etc. (of various shades), the Silver Pencilled is similar to the Partridge.

Plumage, Red Variety, Male and Female: Surface rich, bright, glossy red. Neck hackle of medium shade to match body colour, with a black stripe down the centre of each feather at the lower part. Tail and coverts green-black. Wing, primaries, inner half black, outer half red; secondaries, inner half dark slate or black, outer half to match body. Undercolour dark or slate, clearly defined.

Plumage, White Variety, Male and Female: Pure white, free from yellow or straw tinge.

In both sexes: Beak bright yellow (except in Buff Laced, yellow or horn tipped with yellow; Columbian and Red, yellow or horn; Gold Laced, Partridge, Silver Laced, and Silver Pencilled, horn shading into or tipped with yellow). Eyes bright bay (in the Red, red or orange). Comb, face, wattles and ear-lobes bright red. Legs and feet bright yellow.

Standard Weights: Cock 8 to 9 lb; cockerel $6\frac{1}{2}$ to $7\frac{1}{2}$ lb. Hen 7 lb; pullet $5\frac{1}{2}$ lb.

SCALE OF POINTS

Black Variety

COLOUR (SURFACE 25, UNDERCOLOUR 10)	35
TYPE	25
HEAD	10
SIZE AND CONDITION	15
LEGS	15
	100

Blue Variety

TYPE	25
COLOUR	25
HEAD	15
LEGS	15
SIZE	10
CONDITION	10
	100

Serious Defects: These include black legs devoid of yellow.

Blue Laced Variety

	Male	Female
COLOUR AND MARKINGS	64	73
HEAD	19	10
SIZE AND CONDITION	12	12
LEGS	5	5
	100	**100**

Serious Defects: These include white in tail.

Buff Variety

BACK 10, BODY 12, WINGS 10, TAIL 8	40
HEAD 6, COMB 8, EAR-LOBES AND WATTLES 8	22
SIZE AND CONDITION	20
NECK	10
LEGS	8
	100

Buff Laced Variety

COLOUR AND MARKINGS	70
HEAD	14
SIZE AND CONDITION	11
LEGS	5
	100

Serious Defects: These include black in tail, or excess of blue or grey in lacing.

Columbian Variety

TYPE OR SHAPE	25
COMB	10
EYE	5
BODY COLOUR	15
HACKLE, INCLUDING SCANTINESS	10
TAIL	5
FLIGHTS	5
LEGS	5
TEXTURE	5
CONDITION (INCLUDING ACTIVITY)	7
SIZE	8
	100

Serious Defects (which should be heavily penalised): These include badly crooked breast bone, coarseness. Inactivity. Excess of feather. Overhanging eyebrows. Crooked toes. Brown undercolour. Green eyes.

Gold or Silver Laced Varieties

COMB AND HEAD	10
EAR-LOBES AND WATTLES	4
NECK	8
BREAST AND THIGHS	12
BACK	12
TAIL	6
WINGS	12
FLUFF	5
LEGS	5
SIZE AND CONDITION	14
SHAPE	12
	100

Serious Defects: These include white in tail, or any conspicuous spotting or peppering of ground colour.

Partridge Variety

Male

COLOUR AND MARKINGS (COLOUR OF HACKLES 8, STRIPING OF HACKLES 8, TOP COLOUR 8, BREAST 7, FLIGHTS 5, TAIL 4, UNDERCOLOUR 4, FLUFF 4)...... 48

HEAD (COMB 7, EYES 5, LOBES AND WATTLES 4)............................. 16

TYPE................................. 22

SIZE AND CONDITION.................. 8

LEGS.................................. 6

____

100

Female

COLOUR AND MARKINGS (GROUND COLOUR 13, FORMATION—BREADTH AND FORM OF BLACK MARKS—OF PENCILLING 11, CLEARNESS OF PENCILLING 10, FLUFF 5, HACKLES 4) 43

HEAD (COMB 6, EYES 5, LOBES 4)........ 15

TYPE................................. 22

LEGS................................. 10

SIZE AND CONDITION.................. 10

____

100

Serious Defects: These include slipped wings, wall eyes or eyes that do not match.

Red Variety

INDICATIONS OF EGG PRODUCING MERITS 15

INDICATIONS OF REPRODUCTIVE MERITS.. 15

TYPE AND CARRIAGE.................... 20

COLOUR AND MARKINGS................. 20

HEAD (INCLUDING EYES)............... 10

LEGS................................ 10

CONDITION........................... 10

____

100

Serious Defects: These include shanks other than yellow (allowance made for adult birds and heavy laying females). Coarseness, superfine bone. Any points against egg production, reproduction, or stamina values. Absence of any dark undercolour, and any deformity for which a bird may be passed.

183

Silver Pencilled Variety   Male   Female

	Male	Female
COLOUR AND MARKINGS	48	43
HEAD	16	15
SIZE AND CONDITION	8	10
LEGS	6	10
TYPE	22	22
	100	100

The division of points in the Silver Pencilled is precisely the same as in the Partridge.

White Variety

TYPE	25
COLOUR	25
HEAD	15
LEGS AND FEET	10
SIZE	15
CONDITION	10
	100

Serious Defects (for which a bird may be passed): These include feathers other than white in colour. Coarseness and "Orpington" type.

In all varieties of Wyandottes the following are also regarded as serious defects: Any feathers on shanks or toes. Permanent white or yellow in ear-lobe covering more than one-third of its surface. Comb other than rose, or falling over one side, or so large as to obstruct the sight. Shanks other than yellow, except in mature birds, which may shade to light straw. Any deformity.

YOKOHAMAS

ORIGIN: Japanese
CLASSIFICATION: Light
EGG COLOUR: Tinted

On the Continent of Europe, two breeds of Japanese origin, with long tails are standardised:

1. The Yokohama, with a walnut comb, found in two colours, Red Saddle and White.

2. The Phoenix, with a single comb, found in " game " colours such as Black-Red and Duckwing.

In Britain, both these breeds have been known, shown and standardised under the name of Yokohama. (For photograph see page 205.)

GENERAL CHARACTERISTICS: MALE

Carriage: Stylish and pheasant-like.

Head: Skull small but inclined to be long and tapering. Beak strong and curved. Eyes bright and full of life. Comb single or walnut (in the Red Saddle, walnut only allowed), small and even. Face of fine texture. Ear-lobes small, oval and almond shape, fitting closely. Wattles round and small.

Neck: Long and furnished with flowing hackle.

Body: Fairly long and deep, full round breast, long back tapering to tail, long wings carried rather low but close to the sides. Tail as long and flowing as possible, with a great abundance of side hangers, the sickle and coverts narrow and hard, and the whole tail forming a graceful curve and carried somewhat low.

Legs: Of medium length, the shanks fine and free of feathers. Toes (four) well spread.

FEMALE

The general characteristics are similar to those of the male, allowing for the natural sexual differences.

COLOUR

Black-Red, Duckwings (Silver and Gold), Blue, Red: Beak horn. Eyes ruby-red. Comb, face and wattles bright red. Ear-lobes white or red. Legs and feet yellow, willow or slate-blue.

Plumage: As in the corresponding colours in Game fowl.

Red Saddle: Beak yellow. Eyes urby red. Comb, face, wattles and ear-lobes bright red. Legs and feet bright yellow.

Plumage: White and red. Breast and thighs (red in the male and red-buff in the female), with distinct white spangling at the end of each feather. Male's back and wing bow crimson red, the former vignetted into the Saddle. Remainder white.

White: Beak, legs, and feet white or yellow. Eyes bright red. Comb, face, wattles and ear-lobes, red.

Plumage: Snow-white, free of any straw tinge.

Standard Weights: Male 4 lb to 6 lb. Female $2\frac{1}{2}$ lb to 4 lb.

SCALE OF POINTS

PLUMAGE (QUALITY AND LENGTH OF TAIL 25, AND OF NECK AND SADDLE HACKLES 20)	45
TYPE AND CONDITION	25
HEAD	10
COLOUR	10
LEGS	5
SIZE	5
	100

Serious Defects: Yellow or straw feathers in the White, white in face; other than four toes. Any deformity.

AUTOSEXING BREEDS

An autosexing breed is one in which the chicks at hatching can be sexed by their down colouring. It was when crossing the Gold Campine with the Barred Rock in 1929 that Professor R. C. Punnett and Mr. M. S. Pease discovered the basic principle in their experimental work at Cambridge, and made the Cambar.

Barring is sex-linked, there being a double dose in the male and a single dose in the female, the barring being indicated by the light patch on the head of the chick. This light patch is very similar in chicks of both sexes having black down, but when the barring is transferred to a brown down there is a marked difference. The light head spot on the female chick (one dose) is small and defined, while on the male chick (double dose) it spreads over the body. For that reason, the down colouring in the day-old cockerel is much paler, and the pattern of markings more blurred, than in the newly hatched pullet chick, which has the sharper pattern of markings.

Standards which have been accepted so far by The Poultry Club include Gold and Silver Brussbar; Brockbar; Gold, Silver and Cream Legbar; Gold and Silver Cambar; Gold and Silver Dorbar; Rhodebar and Silver Welbar.

BROCKBAR

ORIGIN: British
CLASSIFICATION: Heavy
EGG COLOUR: Brown

GENERAL CHARACTERISTICS: MALE

Carriage: Upright and graceful.

Type: Body large, deep and compact. Back broad. Breast well rounded. Wings carried well up. Tail rather small, rising from the saddle; sickles gracefully curved.

Head: Strong and refined. Beak short and stout. Eyes bright and prominent. Comb single, medium size, erect, evenly serrated, and free from side sprigs. Ear-lobes and wattles fine, evenly shaped, and well proportioned (i.e., not exaggerated).

Neck: Of medium length, thick and well covered with hackle feathers.

Legs and Feet: Legs wide apart and stout. Thighs not more than three inches long, stiltiness to be avoided. Toes, four, strong, straight, well spread and clean.

Plumage: Close and tight, not fluffy.

FEMALE

The general characteristics are similar to those of the male, allowing for the natural sexual differences.

COLOUR

Plumage, Male and Female: Even pale buffish all over, any shade from pale lemon to orange, but avoiding any suggestion of a red or rusty tinge. As free as possible from black markings or sootiness, and free also from white feathers. The male's plumage is paler than that of the female, though it should never be white. His feathers should be clearly and evenly barred throughout their length. In the female, the barring may be much less distinct, but the clearer the better; and some barring *must* be visible.

In both sexes: Beak, legs and feet yellow or willow. Eyes red or orange. Comb, face, ear-lobes and wattles red.

Standard Weights: Cock 9 lb; cockerel 8 lb. Female 7 lb.

SCALE OF POINTS

TYPE	25
COLOURING AND FEATHERING	10
BARRING	10
LEGS AND FEET	10
CONDITION	10
SIZE AND WEIGHT	15
HEAD	5
HANDLING AND CAPACITY	15
	100

Serious Defects: Much black in feathering. Legs other than yellow or willow; feathering on legs and feet. Other than four toes. White in ear-lobes or wattles.

Downs, Female: Rich salmon buffish, even shade all over, free from sootiness. **Male:** Cream, tinged salmon, even shade all over, free from sootiness.

BRUSSBAR

ORIGIN: British
CLASSIFICATION: Heavy
EGG COLOUR: Tinted or light brown

GENERAL CHARACTERISTICS: MALE

Carriage: Graceful, vigorous and well balanced.

Type: Body oblong-shaped, deep, broad and long. Back broad, long and flat, wide across the shoulders. Breast broad, full and well rounded; breast bone long, straight and well fleshed. Wings of medium length. Tail of medium size and carried at an angle of 45 degrees to the body.

Head: Of medium size and fine quality. Beak short, strong and deep. Comb single, of medium size, firm, upright, and fitting close to the head. Eyes prominent, full and bright, pupils clearly defined. Face smooth and free from coarseness. Wattles and ear-lobes of medium size and fine in texture.

Neck: Fairly long, profusely covered with hackle feathers.

Legs and Feet: Legs of medium length and set well apart. Thighs short, stout and well fleshed. Toes, four, long, straight and well spread.

Plumage: Of light silky texture, free from coarseness.

Flesh: Fine in texture.

Handling: Firm and compact with plenty of muscle.

FEMALE

The general characteristics are similar to those of the male, allowing for the natural sexual differences.

COLOUR

Plumage, Gold Variety, Male: Neck hackle dark gold, sparsely barred with dark grey, sooty generally, feathers gold tipped. Back and shoulder coverts dark gold-grey barred, rich brownish tinge, feathers gold tipped. Wing, bow dark gold-grey barred, gold predominating; primaries dark gold-grey barred; secondaries dark gold-grey barred, upper web with irregular gold markings. Saddle hackles dark gold barred, rich lustrous colour, feathers gold tipped. Breast black, barred, free from salmon colour. Tail evenly barred black or grey, sickles pale, but not white.

Plumage, Female: Head and neck hackles dark gold, softly barred with dark grey. Breast salmon, clearly defined in outline. Body dark grey-gold, with indistinct broad soft barring, feathers showing pale gold shaft and edging. Wing coverts dark grey-gold, barring soft and indistinct; primaries dark grey-gold, barring broad and soft; secondaries dark grey-gold, barring soft and broad, upper web with irregular gold mottling or peppering. Tail and saddle hackles dark grey-gold, softly barred. Tail feathers on the whole darker.

Plumage, Silver Variety, Male: Neck hackle dark silver, sparsely barred with dark grey, but it will be noted that the tip of the hackle feather fades off to pure silver. Back and shoulder coverts silver, with dark grey barring, the feathers being silver tipped, as far as possible free from chestnut. Wing, bow dark grey with silver-grey barring, as far as possible free from chestnut; primaries dark grey barred, some white permissible; secondaries dark grey barred, with the upper web white or grey-mottled. Saddle hackles silver barred with dark grey, the feathers being silver tipped. Breast evenly barred dark grey and silver-grey with well defined outline. Tail evenly barred dark grey and silver-grey, sickles paler. Tail coverts evenly barred dark grey and silver-grey.

Plumage, Female: Head and hackle dark silver with black striping, softly barred grey. Breast salmon, clearly defined. Body dark grey with indistinct broad soft barring, the individual feathers showing lighter shaft and edging. Wing, bow dark grey, as free from chestnut as possible; primaries dark grey, as free from white as possible; secondaries dark grey, upper web lighter grey-mottled or peppered. Tail dark grey with indistinct soft barring.

In both sexes: Beak white or horn. Eyes orange or red. Comb, face and ear-lobes red. Legs and feet white. Skin and flesh white.

Standard Weights: Cock 9 lb; cockerel 8 lb. Hen 7 lb; pullet 6 lb. (All min.).

SCALE OF POINTS

TYPE	30
COLOUR	20
LEGS	10
CONDITION	10
HEAD	20
WEIGHT	10
	100

Cream Legbar Male
Silver Brussbar
Female

Serious Defects: White in ear-lobe.   Squirrel or wry tail.   Feathers on legs and toes.

Defects (for which a bird should be passed): Side sprigs on comb. Eye pupil other than round and clearly defined.   Crooked breast or any bodily deformity.   In the Gold female, excessive chestnut colour on the back or in the plumage generally.   In the Silver, gold feathers as distinct from chestnut.

Downs: As for Legbars in both Gold and Silver.

CAMBAR

Origin: British
Classification: Light
Egg Colour: Tinted

GENERAL CHARACTERISTICS: MALE

Carriage: Upright and alert, with range of body.   A medium type breed.

Type: Body large, deep, broad at shoulders, compact. Back rather long, flat, narrowing slightly to saddle with moderate slope. Breast well rounded, breast bone long and straight.   Wings of medium size and neatly tucked into the body.   Tail medium length, carried well out at 45 degrees to line of the body with nicely curved sickles; sickles and coverts broad and plentiful.

Head: Fine, deep and inclined to width.   Beak short and stout. Eyes large and bright.   Comb single, of medium size, erect, evenly and deeply serrated, free from side sprigs.   Face smooth.   Ear-lobes of medium size and fine texture.   Wattles of medium size, fine and evenly matched.

Neck: Moderately long, profusely covered with feathers.

Legs and Feet: Legs moderately long to hock, free from feathers. Thighs two to three inches long only.   Toes, four, strong, well spread and straight; stiltiness a fault.

Plumage: Tight and silky in texture.

Handling: Firm, with abundance of muscle, showing high qualities expected in a dual-purpose utility breed.

FEMALE

The general characteristics are similar to those of the male, allowing for the natural sexual differences. Comb erect, not falling to one side, tail carried well out.

COLOUR

Plumage, Gold Variety, Male: As regular grey and gold barred as possible all over, except on the head, neck and saddle hackles, which should be gold, and as free from sootiness as possible. The gold as rich in tint as possible.

Plumage, Female: Generally darker than the male, gold on head, neck and saddle hackles rich and clear.

Plumage, Silver Variety, Male and Female: Same as for the Gold, but for " gold " read " silver ".

In both sexes: Beak white. Eyes rich bay. Ear-lobes white. Comb, face and wattles bright red. Legs and feet white.

Standard Weights: Cock 8 lb; cockerel 7 lb. Hen 5½ lb; pullet 5 lb.

SCALE OF POINTS

TYPE AND CARRIAGE	25
COLOUR	20
DUAL-PURPOSE QUALITIES	10
HEAD	15
SIZE AND SYMMETRY	10
CONDITION	10
LEGS AND FEET	10
	100

Serious Defects: Narrowness at shoulders. Light weight. Stiltiness. Falling comb; exaggerated comb, lobes or wattles. Feathers on shanks or feet. Loose feathering. Coarseness. Any point against utility values.

Defects (for which a bird may be passed): Eye pupils other than round and clearly defined. Wry or squirrel tail. Crooked breast bone. Any bodily deformity. Gold feathers in the male of the Silver variety.

Downs, Female (Gold): Mottled chocolate brown on rich gold background, the mottling to be sharply defined in outline. Over the head and eyes there is thin black striping, breaking up into pin-point black spots, giving the characteristic spotty appearance. White markings on the rump to be avoided. Light head patch showing up brightly. **Male:** Much paler, washed out, pattern completely blurred without any well-defined light head patch.

193

DORBAR

ORIGIN: British
CLASSIFICATION: Heavy
EGG COLOUR: Tinted or white

GENERAL CHARACTERISTICS: MALE

Carriage: Upright and alert.

Type: Body large, deep and moderately long. Back broad, saddle feathers of medium length and abundant. Breast broad and well rounded; keel bone long and straight. Wings moderately large and carried well up. Tail of medium size and carried well out, with well curved sickles.

Head: Strong, deep and inclined to width. Beak stout. Eyes large and bright. Comb single, medium to large, evenly and deeply serrated, free from thumb marks or side sprigs. Face smooth. Ear-lobes moderately developed and pendent. Wattles of medium size and fine texture, evenly matched.

Neck: Rather short, tapering, abundant hackle feathers flowing well over the shoulders.

Legs and Feet: Legs wide apart with moderately long hocks, free from feathers. Thighs short and well developed. Toes, five, strong, hard, straight and well spread.

Plumage: Tight.

FEMALE

The general characteristics are similar to those of the male, allowing for the natural sexual differences. Comb, erect or lopped.

COLOUR

Plumage, Gold Variety, Male: Barred all over head and neck. Breast black barred. Wing coverts and saddle hackles gold, all as free from black markings as possible. Flights and tail grey-gold barred. In general the gold parts to be as rich in tint as possible.

Plumage, Female: Body colour silver-grey ground, softly barred. Hackle gold barred. Breast rich, even salmon-pink with sharply defined edges.

Silver Legbar Male
Gold Dorbar Female

Plumage, Silver Variety, Male and Female: As for the Gold. reading " silver " instead of " gold ". Gold feathers in the male to disqualify. Plumage of the male as far as possible free from chestnut smudges.

In both sexes: Beak white. Eyes red or rich bay. Comb, face, ear-lobes and wattles bright red. Legs and feet flesh colour.

Standard Weights: Cock 8 lb (min.); cockerel $6\frac{1}{2}$ lb (min.). Hen 6 lb (min.); pullet about 5 lb.

SCALE OF POINTS

TYPE	30
COLOUR	20
HEAD	20
LEGS	10
WEIGHT	10
CONDITION	10
	100

Serious Defects (to lose points): Narrow back or breast. Light weight. Side sprigs or thumb marks on comb and falling comb in the male. Exaggerated comb, lobes or wattles. Feathers on shanks or toes. Loose feathering.

Disqualifications: Any deformity. Gold feathers in male of Silver variety.

Downs, Female and Male (Gold): As for Gold Legbar.

Downs, Female and Male (Silver): As for the Gold, except that for " brown stripe " read " silver-grey ", and for " the ground colour should be dark brown, though distinctly paler than the stripe " read " ground colour silver-grey ".

LEGBAR

ORIGIN: British
CLASSIFICATION: Light
EGG COLOUR: White or Cream

GENERAL CHARACTERISTICS: MALE

Carriage: Very sprightly and alert, with no suggestion of stiltiness.

Type: Body wedge shaped, wide at the shoulders and narrowing slightly to root of tail. Back long, flat and sloping slightly to the tail. Breast prominent, and breast bone straight. Wings large, carried tightly and well tucked up. Tail moderately full at an angle of 45 degrees from the line of the back.

Head: Fine. Beak stout, point clear of the front of the comb. Eyes prominent. Comb single, perfectly straight and erect, large but not overgrown, deeply and evenly serrated (5 to 7 spikes broad at the base), extending well beyond back of the head and following, without touching, the line of the head, free from " thumb marks " or side spikes. Face smooth. Ear-lobes well developed, pendent, smooth and free from folds, equally matched in size and shape. Wattles long and thin.

Neck: Long and profusely covered with feathers.

Legs and Feet: Legs moderately long. Shanks strong, round and free of feathers. Flat shins objectionable. Toes, four, long, straight and well spread.

Plumage: Of silky texture, free from coarse or excessive feather.

Handling: Firm, with abundance of muscle.

FEMALE

The general characteristics are similar to those of the male, allowing for the natural sexual differences, except that the comb may be erect or falling gracefully over either side of the face without obstructing the eyesight, and the tail should be carried closely and not at such a high angle.

COLOUR

Plumage, Gold Variety, Male: Neck hackle pale straw, sparsely barred with gold and black. Back, shoulder coverts and wing bow pale straw barred with bright gold-brown. Wing coverts (or wing bar) dark grey barred (as in the Cuckoo Leghorn); primaries and secondaries dark grey barred, intermixed with white, upper web of secondaries also intermixed with chestnut. Saddle hackle pale straw barred with bright gold-brown, as far as possible without black. Breast, underparts, dark grey barred (as in the Cuckoo Leghorn). Tail grey barred, sickles paler. Tail coverts grey barred.

Plumage, Female: Hackle pale gold, marked with black bars. Breast salmon, clearly defined. Body dark smoky or slaty grey-brown with indistinct broad soft barrings, the individual feather showing paler shaft and slightly paler edging. Wings dark grey-brown. Tail dark grey-black with slight indication of lighter broad bars.

Plumage, Silver Variety, Male: Neck hackle silver, sparsely barred with dark grey but tips of feathers fade off to pure silver. Saddle hackle silver, barred with dark grey, the feathers tipped with silver. Back and shoulder coverts silver, with dark grey barring, the feathers tipped with silver. Wing bow dark grey with silver-grey barring; primaries dark grey, some white permissible ; secondaries dark grey with tips of upper web white. Breast evenly barred dark grey and silver-grey, with well defined outline. Tail and tail coverts evenly barred dark grey and silver-grey, sickles being paler.

Plumage, Female. Head and neck hackle silver, with black striping, softly barred grey. Breast salmon, clearly defined. Body silver-grey, with indistinct broad soft barring, individual feathers showing lighter shaft and edging. Wings silver-grey, as free from chestnut as possible; primaries silver-grey, as free from white as possible; secondaries silver-grey, upper web a lighter grey mottled. Tail silver-grey with indistinct soft barring.

In both sexes: Beak yellow or horn. Eyes orange or red, pupils clearly defined. Comb, face and wattles bright red. Ear-lobes pure opaque white (resembling white kid) or cream, the former preferred. Slight pink markings and pink edging not unduly to handicap an otherwise good bird for utility purposes. Legs and feet yellow, orange or light willow in the female.

Plumage, Cream Variety, Male: Neck hackles cream, sparsely barred. Saddle hackles cream, barred with dark grey, tipped with cream. Back and shoulders cream with dark grey barring, some chestnut permissible. Wings, primaries dark grey, faintly barred, some white permissible; secondaries dark grey more clearly marked; coverts grey barred, tips cream, some chestnut smudges permissible.

Breast evenly barred dark grey, well defined outline. Tail evenly barred grey, sickles being paler, some white feather permissible. Crest cream and grey, some chestnut permissible.

Plumage, Female: Neck hackles cream, softly barred grey. Breast salmon, well defined in outline. Body silver grey, with rather indistinct broad soft barring. Wings, primaries grey-peppered; secondaries very faintly barred; coverts silver grey. Tail silver grey, faintly barred. Crest cream and grey, some chestnut permissible.

In both sexes: Beak yellow. Eyes orange or red. Comb, face, and wattles red. Ear-lobes pure opaque, white or cream, slight pink markings not unduly to handicap an otherwise good male. Legs and feet yellow.

Note: This is a crested variety, laying a blue, green or olive egg.

Standard Weights: Cock 7 to $7\frac{1}{2}$ lb; cockerel 6 to $6\frac{1}{2}$ lb. Hen 5 to 6 lb; pullet $4\frac{1}{2}$ to 5 lb.

SCALE OF POINTS

TYPE	30
COLOUR	20
HEAD	20
LEGS	10
CONDITION	10
WEIGHT	10
	100

Serious Defects: Male's comb twisted or falling over. Ear-lobes wholly red. Any white in face. Legs other than orange, yellow or light willow. Squirrel tail.

Defects (for which a bird may be passed): Side sprigs on comb. Eye pupil other than round and clearly defined. Crooked breast. Wry tail. Any bodily deformity.

Downs, Female (Gold): Brown stripe type. The stripe should be broad and very dark brown, extending over the head, neck and rump. The edges of the stripe should be clearly defined rather than blurred and blending with the ground colour—the sharper the contrast, especially over the rump, the better. The ground colour should be dark brown, though distinctly paler than the stripe. A pale ground colour and a narrow or discontinuous stripe are to be avoided. A light head spot should be visible, though usually it is small. It should be well defined in outline and should show up as clearly as possible against the brown background. **Male:** The down is much paler in shade, the pattern being blurred and washed out from head to rump.

Downs, Female (Silver): Silver-grey type. The stripe should be very dark brown, extending over the head, neck and rump. The edges of the stripe should be clearly defined, not blurred and blending with the ground colour—the sharper the contrast, especially over the rump, the better. The stripe should be broad; a narrow or discontinuous stripe should be avoided. A light head patch should be visible, clearly defined in outline, showing up brightly against the dark background. **Male:** The down is much paler in tint, the pattern being blurred and washed out from head to rump; it may best be described as pale silvery-slaty.

Downs (Cream): As silver.

RHODEBAR

ORIGIN: British
CLASSIFICATION: Heavy
EGG COLOUR: Brown

GENERAL CHARACTERISTICS: MALE

Carriage: Upright and graceful.

Type: Body large, fairly deep, broad and long. Back broad, long and somewhat horizontal in outline. Breast broad, full and well rounded. Wings carried well up, the bows and tips covered by the breast feathers and saddle hackle. Tail rather small, rising slightly from the saddle, the sickle of medium length, well spread and nicely curved, the coverts being sufficiently abundant to cover the stiff feathers.

Head: Strong, but not thick. Beak moderately curved, short and stout. Eyes large and bright. Comb single, medium size, straight, upright, well set on, with well-defined serrations, and free from side sprigs. Face smooth. Ear-lobes of fine texture, well developed and pendent. Wattles to correspond with size of comb and moderately rounded.

Neck: Of medium length and profusely covered with feathers flowing over the shoulders, but not too loosely carried.

Legs and Feet: Legs wide apart and of medium length, stout and strong and free from feathers. Thighs large with well-rounded shanks of medium length. Toes, four, strong, straight and well spread.

Handling: Firm, with abundance of muscle.

Plumage: Of silky texture, free from coarse or excessive feather.

FEMALE

The general characteristics are similar to those of the male, allowing for the natural sexual differences.

COLOUR

Plumage, Male: Hackle deep red-gold barred, with centres black and grey-white barred, the black centre portions rather longer than the grey-white; the front of the cape showing less black, the feathers towards the tips of the cape lying on the back showing wider black and grey-white barring. Wing primaries, lower web red-gold, faintly barred, upper grey and white barred, slightly gold tinted; secondaries, the whole alternately black, white and gold barred, lower web showing more gold; flight coverts very bright red-gold and white barred, tips red-gold. Wing bows very brilliant chestnut red and gold barred. Tail, including sickles, uniform black and white barring from tip to base, including the shaft. Tips black. Saddle hackle deep red-gold and grey-white and narrower black barring towards the tips. Back and saddle deep red-gold barred, with occasional black bars towards the end of the feathers. Under-colour light creamy buff. Breast uniformly barred, deep red-gold and creamy white and black.

Plumage, Female: Hackle deep buff-red with bright chestnut edges, each feather with deep buff, gold, black and white narrow barring, the barring becoming narrower as it approaches the lower cape feathers. Tail feathers black with reddish tinge. Wing primaries, upper web red-buff, lower black; secondaries buff-red. Remainder, general surface dark buff-red barred with buff and buff-red, the tips of the feathers of the lighter colour. Undercolour creamy buff-red, as deep as possible. Quills yellow.

In both sexes: Beak red-horn or yellow. Eyes orange or red, pupils clearly defined. Comb, face, ear-lobes and wattles bright red. Legs and feet bright yellow.

Standard Weights: Cock 8½ lb (min.); cockerel 8 lb.   Hen 6½ lb (min.); pullet 5½ lb.

SCALE OF POINTS

TYPE	30
COLOUR.....................................	20
LEGS..	10
CONDITION.................................	15
HEAD.......................................	20
WEIGHT....................................	5
	100

Serious Defects: Male's comb twisted or falling over.   Ear-lobes other than red.   Legs other than yellow, orange or light willow. Squirrel or wry tail.   Side sprigs on the comb.   Eye pupils other than round and clearly defined.   Crooked breast or any bodily deformity.

WELBAR

ORIGIN: British
CLASSIFICATION: Light
EGG COLOUR: Brown

GENERAL CHARACTERISTICS: MALE

Carriage: Upright, alert and active.

Type: Body well built on good constitutional lines.   Back broad and long.   Breast full, well rounded and broad.   Wings moderately long, carried close to side.   Tail fairly large and full, carried high, but not squirrel.   Abdomen long, deep and wide.

Head: Refined.   Beak strong, short and deep.   Eyes large, bright. Comb single, medium size, firm and upright, free from any twists or excess, clear of the nostrils, fine texture, five to seven broad and even serrations, the back following closely, but not touching line of the skull and neck.   Face smooth and without overhanging eyebrows.   Ear-lobes small and almond shaped.   Wattles of medium size, fine texture, close together.

Neck: Fairly long, slender at top, finishing with abundant hackle.

Legs and Feet: Thighs to show clear of the body.   Shanks of medium length and bone, well set apart, free from feathers with soft sinews and free from coarseness. Toes, four, long, straight and well spread out.

Plumage: Tight, silky, free from excess or coarseness and free from bagginess at the thighs.

Handling: Compact, firm and neat in bone throughout.

FEMALE

The general characteristics are similar to those of the male, allowing for the natural sexual differences.

COLOUR

Plumage, Silver Variety, Male: Head silver. Hackles silver, black ticking permissible. Back, shoulders, coverts and wing bow silver. Wing coverts (or bar) black barred; primaries and secondaries, inner web black barred, outer web silver. Tail black barred. Breast black barred with silver mottling.

Plumage, Female: Head and hackle silver with black striping barred with white. Breast salmon. Back and wing bow and bar, light grey, faintly barred, free from salmon smudges; primaries and secondaries, outer web silver, coarsely stippled with dark grey, inner web dark grey and faintly barred. Tail dark grey and faintly barred.

In both sexes: Beak yellow. Eyes red. Comb, face, ear-lobes and wattles bright red. Legs and feet yellow.

Standard Weights: Cock $7\frac{1}{2}$ lb; cockerel $6\frac{1}{2}$ lb. Hen 6 lb; pullet $4\frac{1}{2}$ to 5 lb.

SCALE OF POINTS

TYPE	30
COLOUR	20
HEAD	20
LEGS	10
CONDITION	10
WEIGHT	10
	100

Serious Defects: Side sprigs to the comb. White in lobe. Feathers on the legs, hocks or between the toes. Comb other than single. Other than four toes. Gold feathers on the male. Legs other than yellow. Badly crooked or duck toes. Any bodily deformity. Coarseness, beefiness, and anything that interferes with productiveness and the general utility of the breed.

OTHER BREEDS

Listed below are breeds which in some cases have been known and standardised in Britain, but appear to have died out, though they are still to be seen on the Continent (Crève-Cœur, La Flèche and Orloff). Others are still to be found in Britain but are very rare (Lakenvelder, Sicilian Buttercup).

Crève-Cœur: Origin French. Classification heavy. Egg colour white. Comb horn-type (U-shaped). Face muffled. Body broad, square built with bold upright carriage. Legs clean and short. Toes, four. Colour lustrous green-black. Weight, male 9 lb, female 7 lb.

La Flèche: Origin French. Classification heavy. Egg colour white. Comb two-horned, similar to the Crève-Cœur but has additional two small " studs " just in front of the nostrils. Face red. Ear-lobes white. Body resembles the Spanish but larger. Legs slate coloured. Toes, four. Colour a dense sheeny green-black. Weight, male $8\frac{1}{2}$ lb, female $7\frac{1}{2}$ lb.

Lakenvelder: Origin German. Classification light. Egg colour white. Single comb. Face red. Ear-lobes white. Body long and well rounded with broad and short back. Carriage upright. Legs medium length and clean. Toes, four. Colour black and white. Weight, male 6 lb, female $4\frac{1}{2}$ lb.

Malines: Origin Belgium. Classification heavy. Egg colour pale brown. Single comb. Face red. Body deep, broad with a long flat back. Carriage upright. Legs white, strong, and fairly long. Shanks lightly feathered. Toes, four, outer lightly feathered. Colours, Cuckoo, Blue, Black, Gilded Black, Silvered Black, Gilded Cuckoo, Ermine (similar to Light Brahma) and White. Weight, male 9 lb, female 7 lb.

Marsh Daisy: Origin British. Classification heavy. Egg colour white. Rose comb. Face red. Ear-lobes white. Body fairly broad, square and " blocky " in appearance with well-rounded prominent breast. Legs pale willow-green, moderately long. Toes, four. Colours, Black, Brown, Buff, Wheaten and White. Weight, male $6\frac{1}{2}$ lb, female $5\frac{1}{2}$ lb.

Duckwing Yokohama Male
Sicilian Buttercup Male
Sumatra Game Male

Norfolk Grey: Origin British. Classification heavy. Egg colour dark tint to brown. Single comb. Face red. Body long and broad at shoulder, full and rounded breast carried upwards. Legs black or slate-black, fairly short. Toes, four. Colour silver-white and black. Weight, male 8 lb, female 6 lb.

Orloff: Origin Russian. Classification heavy. Egg colour brown. Raspberry type comb. Face muffled and beetle brows with gloomy and vindictive expression. Body like Indian Game, but long-legged and not so massive. Ear-lobes red. Legs rich yellow. Four varieties, Black, Mahogany, Spangled and White. Weight, male 8 lb, female 6 lb.

Phœnix: Origin Oriental. Classification light. Egg colour white-tinted. Pea comb. Face red. Ear-lobes pure white or pure red. Body fairly long and deep, full rounded breast. Back lengthy, tapering towards tail which is long and as flowing as possible with great abundance of side hangers. Legs yellow, willow or slate-blue according to colour of variety, medium length. Toes, four. Colours, Golden or Black-reds, Duckwings, Spangled and White. Weight, male 6 lb, female 4 lb.

Sicilian Buttercup: Origin Continental. Classification light. Egg colour white. Cup-shaped comb. Comb, face wattles and ear-lobes bright red. Body moderately long and deep, upright carriage. Back broad and sloping towards saddle. Breast full and round. Legs willow-green and of moderate length. Toes, four. Varieties, Brown, Golden, Golden Duckwing, Silver and White. Weight, male 6½ lb, female 5½ lb.

Wherwell: Origin British. Classification heavy. Egg colour tinted. Single comb. Face red, beak white or horn striped with white, eyes yellow to red. Body broad across the shoulders with prominent butts. A compact outline showing good depth of body with rounded sides and wings. Toes, four. Gold and Silver varieties. Weight, male 10 lb, female 8 lb.

White Surrey: Origin British. Classification heavy. Egg colour tinted. Rose comb. Face red, beak white, eyes red or bright orange. Body fairly long with good width and depth. Toes, four. Colour white. Weight, male 10 lb, female 8 lb.

BANTAM STANDARDS

 THIS section includes all breeds of bantams standardised or accepted in Britain. Standards of specialist bantam breed clubs have been adopted where they exist. Where these are not available, those of large breed specialist clubs catering for bantams have been used; and where neither of these considerations apply, the last published standards of The Poultry Club have been accepted, with such revisions as proved necessary to modernise them and to bring them all into one general form.

Concerning the vexed questions of size and weight there may be some controversy or disagreement. It should be observed, however, that as far as possible all interests have been consulted.

Where accepted bantam views and methods differ from large breeds, remarks, descriptions and advice (in parentheses or otherwise) have been added by the British Bantam Association, with whose authority and approval these standards have been produced.

It will be observed that not every colour is fully described. This is because here and there colours and markings have become known and tacitly accepted without having been officially standardised.

Of true bantams, without counterparts in large fowl, few breeds exist. They include Belgian Bearded, Cochins or Pekins (originally known simply as Pekins and imported from China as genuine bantams over 100 years ago), Booted, Japanese, Nankins, Rose-combs and Sebrights—the last-named evolved in Britain about 150 years ago. To these must be added Old English Game and Modern Game, which like the Rosecomb were developed mainly from the common bantam of the countryside, and by virtue of long development on bantam lines must now be accepted as true bantams.

Other genuine bantams have been developed in Europe but are not standardised here. Breeds with counterparts in large poultry might well be described instead as miniature fowl. No other accepted name or description has yet been applied to them, and no attempt has been made in these standards to segregate them. Indeed, some bantamised varieties of large fowl have been standardised so long, and now contain so much blood from bantam sources, that they may well be grouped among true bantams. This applies to several varieties of Wyandotte, which were among the earliest of the more modern creations; also perhaps to Frizzles and Scots Greys.

The essential differences in breed character, wings, head points and carriage between old-established true bantams and the more modern miniature fowl need not be discussed here. Bantamising large breeds without perpetuating these distinctive bantam characteristics (which may be obnoxious when not desired) is a problem that takes time, even for the cleverest of breeders. The success achieved in some miniature varieties shows that these difficulties can be overcome.

ANCONA

THESE are miniatures of the large breed, and are of Mediterranean character. Weights a little higher than former P.C. Standards have been suggested. The large breed club caters for bantams.

GENERAL CHARACTERISTICS: MALE

Carriage: Active, bold and alert.

Type: Body broad, close and compact. Back moderately long, tapering slightly to saddle. Breast full and broad, carried forward and up. Wings large and well tucked up. Tail full, carried well out, with plentiful furnishings and flowing sickles.

Head: Deep, moderate in length, carried well back. Beak medium length and moderately curved. Eyes bright and prominent. Comb single, upright, medium size, with five to seven broad even serrations forming a regular curve finishing well back, blade following line of head, free from excrescences or thumb marks. Face smooth and fine. Ear-lobes medium size, almond-shaped, free of folds. Wattles long and fine.

Neck: Long, arched, profusely hackled.

Legs and Feet: Legs of medium length, set well apart. Shanks strong and free from feathers, thighs not much seen. Toes, four, rather long and thin, well spread.

FEMALE

With the exception of the comb, which should fall gracefully to one side of the face without obstructing the sight, the general characteristics resemble those of the male, allowing for the natural sexual differences. Abdominal development should also be more marked.

COLOUR

Plumage, Male and Female: Beetle-green with V-shaped white tippings throughout (the more evenly tipped the better) with no inclination to splashes or lacing. White tips to be clear of black or grey streaks. Feathers black to roots.

In both sexes: Beak yellow, shaded with black or horn. A wholly yellow beak is not desirable. Eyes orange-red, with hazel pupil. Comb, face and wattles bright red, face free from white. Ear-lobes white. Legs and feet yellow, mottled with black; the more evenly mottled the better.

Suggested Weights: British Bantam Association—Male 22 to 26 oz. Female 18 to 22 oz. Ancona Club—Male 20 to 24 oz. Female 18 to 22 oz.

Note: A rose combed variety is standardised in the large breed, with comb to resemble the Wyandotte, and this is acceptable in Bantams.

SUGGESTED SCALE OF POINTS

EYES	5
LOBES	5
LEG COLOUR	5
GROUND COLOUR AND UNDERCOLOUR	15
TYPE AND CARRIAGE	15
COMB	10
BEAK COLOUR	5
PURITY OF WHITE AND TIPPING	20
SIZE AND TEXTURE	15
CONDITION	5
	100

Serious Defects: White in face. Light undercolour. Plumage other than black and white. Crooked toes. Knock knees. Squirrel or wry tail. Roach back or other structural deformity to disqualify.

ANDALUSIAN

Since the publication of the last edition of this work Andalusian bantams have added to standards. They should be miniatures of the big breed, Mediterranean in character, very similar to Minorcas but less developed in head points.

GENERAL CHARACTERISTICS: MALE

Carriage: Upright, bold and active.

Type: Body long, with broad shoulders tapering to tail. Breast full, round; close and compact plumage. Wings long and well tucked up, ends hidden under saddle hackles. Tail large and flowing, carried well up but not fan-shaped.

Head: Skull moderately long and fairly wide. Beak stout and of medium length. Eyes prominent. Comb single, upright, medium size, deeply serrated, the spikes broad at base, blade of comb following line of head but not touching neck; no thumb marks or side sprigs. Face smooth. Ear-lobes almond-shaped, flat, undished, medium size, unwrinkled and fitting close to head. Wattles long and fine.

Neck: Long and well hackled.

Legs and Feet: Legs long. Shanks and feet free from feathers. Toes, four, straight, thin and well spread.

FEMALE

With the exception of the comb, which falls with a single fold to one side and partly covers the eye, the general characteristics are similar to those of the male, allowing for the natural sexual differences.

COLOUR

Plumage, Male and Female: Clear blue, each feather crisply and distinctly edged with medium width black lacing, except sickles, which are dark or black. The hackles in both sexes are black with a rich gloss, the female's neck hackles showing broad lacing at the base.

In both sexes: Beak dark slate or horn. Eyes dark red or red-

brown. Comb, face and wattles bright red. Ear-lobes white.
Legs and feet dark slate or black.

Suggested Weights: Male 24 to 28 oz.   Female 20 to 24 oz.

SCALE OF POINTS

GROUND COLOUR	30
COMB	10
LOBES	5
LACING	20
FACE	10
TYPE, CARRIAGE, TAIL AND CONDITION	25
	100

Serious Defects: White in face or red in lobes. White feathers.
Sooty ground colour. Comb not upright in males. Red or yellow
in hackles. Erect comb in females. Any deformity.

AUSTRALORP

THESE are miniatures of the large breed, and the specialist club
caters for their interests.

GENERAL CHARACTERISTICS : MALE

Carriage: Erect and graceful, denoting an active fowl, the head
being carried well above the tail line.

Type: Body deep and broad, showing somewhat greater length
than depth. Back broad across shoulders and the saddle, with a
sweeping curve from neck to tail. Breast full and rounded, carried
well forward without bulging; breast bone long and straight.
Wings compact and carried closely in, the ends being covered by
the saddle hackles. Tail full and compact, rising gradually from
the saddle in an unbroken line; the sickles gracefully curved, but
not long and streaming.

Head: Finely modelled with skull rounded. Beak slightly curved,
strong, of medium length. Eyes, large, prominent and expressive;
high in skull standing out well when viewed from front or back.

Comb single, medium in size, erect, evenly serrated (four to six serrations) and blade tending downwards without touching the neck, texture fine, but not of polished appearance. Face full, fine in texture, clean, free of feathers, wrinkles and overhanging brows. Ear-lobes small and elongated. Wattles medium in size, rounded at bottom and corresponding in texture to comb.

Neck: Fairly long, fine at the junction of the head, with a gradual outward curve to the back, widening distinctly at the shoulders.

Legs and Feet: Legs medium in length, strong, rounded in front, and spaced well apart, the hocks nearly covered by body feathering, and the whole of the shanks showing below the underline. Shanks and feet (four toes) free from feathers or down.

Plumage: Feathering soft but close, with a minimum of fluff, the lower body fluff being only sufficient to cover the thighs.

Skin: Fine in texture.

FEMALE

The general characteristics are similar to those of the male, allowing for the natural sexual differences. The pelvic bones should be pliable, not showing an excess of fat or gristle; the abdominal skin being pliable without an excess of internal fat. All these parts to be of fine texture; any indication of coarseness should be dis-countenanced.

COLOUR

Plumage, Male and Female: Black with lustrous green sheen. In both sexes: Beak black. Eyes black or dark brown iris, black preferred. Face, comb, ear-lobes and wattles bright red. Legs and feet black with white soles. Skin white.

Standard Weights: Male 36 oz (max.). Female 28 oz (max.).

SCALE OF POINTS

TYPE	35
HEAD (EYES 10, FACE 5, SKULL 5, COMB AND WATTLES 5)	25
PLUMAGE (COLOUR, QUALITY AND CHARACTER OF FEATHERING)	12
TEXTURE AND FREEDOM FROM COARSENESS	15
CONDITION	8
LEGS AND FEET	5
	100

Defects (for which birds should be passed): Any deformity such as wry tail, roach back, crooked breast bone, crooked toes, webbed feet. Yellow or willow colour in legs or feet. Yellow or pearl coloured eyes. Feathering on shanks or feet. Side sprigs on comb. Split or twisted wing and slipped wing.

Serious defects: Red, yellow or white in feathers; permanent white in ear-lobes.

BARNEVELDER

THE miniatures are sponsored by the large breed club. They have attained a reasonable degree of perfection and now enjoy classes at many of the leading shows.

GENERAL CHARACTERISTICS: MALE

Carriage: Alert and upright, body compact and back concave.

Type: Body moderately long, deep and broad. Shoulders broad. Saddle high set. Breast deep and full. Stern full. Wings short and carried high. Tail full but not long, with graceful sweep.

Head: Skull neat. Beak short and strong. Eyes bold and prominent. Comb single, upright, medium size, well serrated, firm at base, blade following the neck. Face smooth, free from feathers. Ear-lobes long and narrow. Wattles of medium size.

Neck: Fairly long, profusely feathered and erect.

Legs and Feet: Legs of medium length, fairly stout; no feathers on shanks and feet. Toes, four, well spread.

Plumage: Fairly close and tight and of good texture.

FEMALE

The general characteristics are similar to those of the male, allowing for the natural sexual differences.

COLOUR

Plumage, Male: Neck hackle black with red markings, not regularly striped. Breast black-laced on red-brown ground. Thighs, tail and underparts black. Back deep red with wide black lacing. Wings red-brown with broad lacing of black and red-brown bar;

primaries and secondaries black with red outer edges. Saddle black with red markings. All black markings and feathers to have rich beetle-green sheen.

Plumage, Female: Neck hackles black, with or without brown markings. Breast red-brown with broad black lacings. Back and cushion red-brown, double laced with black, outer lacing very broad. Wings black with red-brown bar; primaries and secondaries black with red-brown outer edges. Tail black, coverts red-brown laced with black.

A Partridge variety also exists but is seldom seen in bantams. The essential difference from the double-laced variety is that the females have red-brown plumage with a glossy black outer lacing only, the remainder of ground colour being evenly stippled with small black peppering.

In both sexes: Beak yellow with dark point. Eyes orange. Comb, face, wattles and ear-lobes red. Legs and feet yellow.

Suggested Weights: Male 32 oz (max.).   Female 26 oz (max.).

SCALE OF POINTS

TYPE AND SIZE	30
PLUMAGE AND TEXTURE	15
HEAD POINTS	10
COLOUR AND MARKINGS	25
LEGS AND FEET	10
CONDITION	10
	100

Serious Defects: White in lobes. Side sprigs on comb. Legs and feet other than yellow. Feathered legs or feet. White in plumage. Any deformity.

BELGIAN BEARDED BANTAMS

THE only varieties of Belgian Bantams yet standardised in Britain are Barbu d'Anvers (Bearded Antwerp) and Barbu d'Uccle (Bearded Uccle). The following standards are adopted by the British Belgian Bantam Club from Belgian National Standards.

Belgian Bearded Bantams are old-established true bantams, without counterparts in large breeds. Each of the two main varieties (d'Anvers and d'Uccle) has many colour variations, some of them intricate, and all attractive.

BARBU d'ANVERS

GENERAL CHARACTERISTICS: MALE
(The d'Anvers is always rose-combed and clean legged.)

Carriage and Appearance: Small, proud, standing bolt upright, with the head thrown well back; proud and provoking (appearing always ready to crow) with characteristic great development of neck hackle.

Type: Body broad and short, with arched breast carried well up. Back very short, slanting downwards to tail. Wings medium length, carried sloping towards ground. Tail carried almost perpendicularly, the main tail feathers strong and not hidden by the narrow sickle feathers; the two largest sickles slightly curved and sword-shaped, the remainder in fan-like tiers to junction with saddle hackle.

Head: Appearing rather large. Beak short, strong and curved, carrying a longitudinal band of light or dark colour in keeping with the plumage. Comb curved, broad in front, ending in a leader or spike at rear; for preference covered with small tooth-like points, or alternatively hollowed and ridged. Point or leader to follow line of neck. Eyes large and prominent, as dark as possible, colour to vary in keeping with plumage. Face covered with relatively long feathers, standing away from the head, sloping backwards and forming whiskers which cover ears and ear-lobes. Brow heavily furnished with feathers. Beard composed of feathers turned horizontally backwards from both sides of the beak and from the centre vertically downwards, the whole forming a collar and giving a muffed effect. (*Note:* In the Belgian National Club Standard the beard is required to form a trilobe, as in the Barbu d'Uccle.) Ear-lobes small, wattles rudimentary only, but preferably none.

Neck: Of moderate length, the hackles thick and convexly arched, entirely covering back and base of neck and forming closely-joined cape at front.

Legs: Thighs short, with medium-length shanks free from feathers. Toes, four, strong and straight, with nails of same colour as the beak.

FEMALE
With certain exceptions the general characteristics are similar to

those of the male, allowing for the natural sexual differences. **Head:** Appearing broader than that of the male and more owl-like. **Neck:** Hackle inclining backwards and forming a ruffle behind the neck, with feathers broader and more developed than in the male. The female hackle, contrary to that of the male, diminishes in thickness towards bottom of neck. **Tail:** Short, carried sloping upwards, slightly curved towards the end and a little open. **Carriage and Appearance:** A little bird, compact, plump, very lively, with characteristically full rounded neck hackle and well-developed whiskers.

Size: As small as possible. The British Belgian Bantam Club does not advocate a weight standard for the breed, but purely as a general guide, suggests: Male 24 to 28 oz, Female 20 to 24 oz; with usual variations for age and maturity. These should be taken as maximum weights.

SCALE OF POINTS

HEAD, COLLAR, BEARD AND NECK	25
COMB	5
COLOUR	15
SIZE	15
BEAK AND NAILS	5
LEGS AND FEET	5
CARRIAGE	10
GENERAL APPEARANCE	20
	100

Serious Defects: Wattles strongly developed. Conspicuous ear-lobes. Squirrel or wry tail. Excessive length of leg.

Disqualifications: Any trace of faking. Wattles cut or removed. Single comb. Absence of beard or whiskers. Feathers on shanks or feet. More than four toes. Yellow colouring of legs, feet or skin.

BARBU d'UCCLE

GENERAL CHARACTERISTICS: MALE

(The d'Uccle is always single-combed and feather-legged.)

Carriage and Appearance: Typically male with a majestic manner, short and broad, with characteristic heavy development of plumage.

Type: Body broad and deep. Back very broad, almost hidden by enormous neck hackle. Breast extremely broad, the upper part very developed and carried forward, the lower part resembling a breast plate. Wings close, fitting tight to body, sloping downwards and incurved towards but not beyond the abdomen; wing butts covered by neck hackle and tips (or ends of flights) covered by saddle hackle, which should be abundant and long. Tail well furnished, close and carried almost perpendicularly to line of back, the two main sickles slightly curved, the remainder in regular tiers and fan-like down to junction with saddle hackle.

Head: Slender and small, with a longitudinal depression towards the neck. Beak short and slightly curved. Comb single, fine, upright, less than average size, evenly serrated, rounded in outline, blade following line of neck. Eyes round, surrounded with bare skin. Brow heavily covered with feathers becoming gradually longer towards the rear, with a tendency to join behind the neck. Beard as full and developed as possible, composed of long feathers turned horizontally from the two sides of beak, and vertically under the beak downwards, the whole forming three ovals in a triangular group. Ear-lobes inconspicuous. Wattles as small as possible.

Neck: Furnished with silky feathers starting behind the beard at sides of throat, with a tendency to join behind the neck to form a mane. Hackle very thick and convexly arched, reaching to shoulder and saddle and covering the whole back.

Legs and Feet: Legs strong, short and well apart, the hocks having clusters of long stiff feathers close together, starting from the lower outer thigh, inclined downwards and following outline of wings. Front and outside of shanks must be covered with feathers, short at top of shanks and gradually increasing in length towards the foot feather; footings turned outwards horizontally, with ends slightly curved backwards. Outer toe and outside of middle toe covered with feathers similar to shank feather.

FEMALE

With certain exceptions the general characteristics are similar to those of the male, allowing for the natural sexual differences. **Beard:** Resembling that of the male but formed with softer and more open feathers. **Neck:** Hackles very thick and convexly arched, composed of broad and rounded feathers, the shape of the mane resembling that of the male. **Tail:** Short, flat in width and not high, the lower main feathers diminishing evenly in length. **Carriage and Appearance:** A quiet little bird, short, thick and cobby.

Size: Dwarf, as small as possible. The British Belgian Bantam Club does not advocate a weight standard, but purely as a general guide, suggests: Male 28 to 32 oz, Female 24 to 28 oz; with usual variations for age and maturity. These should be regarded as maximum weights.

SCALE OF POINTS

HEAD, COMB AND BEAK	10
BEARD	15
NECK HACKLE	10
FEET AND HOCKS	15
WINGS AND TAIL	10
COLOUR	15
SIZE	10
GENERAL APPEARANCE	15
	100

Serious Defects: Strongly developed wattles. Conspicuous ear-lobes. Squirrel or wry tail. Excessive length of leg.

Disqualifications: Any trace of faking. Wattles cut or removed. Comb other than single. Absence of beard or whiskers. Poorly feathered shanks or feet. More than four toes. Yellow legs, feet or skin.

COLOUR

Main colours only are fully described. Belgian Bantams exist in an extraordinary choice of colours, probably unequalled in any other breed, and much too numerous to be given in detail.

Comb, ear-lobes and rudimentary wattles are red in all colour-varieties.

Plumage, Millefleurs, Male: This is a very intricate and attractive colour scheme. Briefly the head is orange-red with white points. The beard is of black feathers laced with very light chamois, each feather ending with a round black spot with a white triangular tip. Neck hackle black with golden shafts, and broadly bordered with orange-red, each feather having a black end tipped with a white point. The extraordinary abundance of neck hackle makes the main colour appear wholly orange-red, the black parts being scarcely visible. Back red, shading to orange towards saddle hackle. Wing bows mahogany-red, each feather tipped with white. Wing bars russet-red with lustrous green-black pea-shaped spots

Blue D'Anvers Male
Millefleurs D'Uccle
Female

at ends, finishing with silvery-white triangular tips, the whole forming regular bars across the wings. Primaries black with a thin edging of chamois on outside. The visible lower third of each secondary feather chamois, upper two-thirds black. Remainder of wing a uniform chamois, each feather having at end a large pea-shaped white spot on a black triangle, the tips spaced evenly to conform with shape and outline of wing. (Note the reversal of these pattern-markings from the normal arrangement.) Tail feathers black with a metallic green lustre, having a fine edging or lacing of dark chamois, and terminating with a white triangle. Breast, foot-feathering and remainder of plumage throughout of golden chamois ground colour, each feather having a light chamois shaft and finished with a black pea-shaped spot tipped with a white triangle.

Plumage, Female: Ground colour uniform golden chamois, each feather terminating with a black pea-shaped spot tipped with a white triangle. Tail feathers black, finely laced with chamois and with white tips. Wing markings and other plumage as described for male, allowing for natural sexual differences.

In both sexes: Eyes orange-red with black pupils. Beak and nails slate-blue. Legs and feet slate-blue.

Defects to be avoided: Ground colour too light or washed out. White markings excessively gay or unevenly distributed.

Plumage, Porcelaine, Male and Female: This is an extraordinarily delicate colour pattern. Markings and patterns generally are as described for Millefleurs in both sexes, with the exception that ground colour is light straw and the pea-shaped spots are pale blue, tipped with white triangles. Pale blue is substituted for the black of the Millefleurs in both sexes.

In both sexes: Eyes orange-red with black pupils. Beak and nails slate-blue. Legs and feet slate-blue.

Defects to be avoided: Ground colour too light or washed out. White markings too gay or unevenly distributed.

Plumage, Quail, Male: This is a very striking colour scheme, with head feathers dark green-black, finely laced with gold. Beard golden-buff or nankin, shading darker towards the eyes, where plumage is black, finely laced with gold. Neck hackle with brilliant black ground, sharply laced with buff, having a golden lustre and yellowish-buff shafts. Back, black ground colour with gold lacing, starting in middle of feathers and narrowing towards the tips, forming lance-like points with golden-silky barbs and well-defined light-ochre coloured shafts from root to point. These feathers are relatively broad under the neck, but narrower and longer nearing the saddle hackle. Colour more intense and black ground more pronounced towards the saddle hackle. Wing bows light gold, lower half of each feather black and clearly defined from the upper half, which should be nankin. Wing bars light ochre, each feather having a black triangular tip, the triangles forming two regular bars across the wing. Bottom third of secondaries chamois-colour, other two-thirds dull black. Primaries dull black, hidden when the wing is closed. Tail black with metallic green lustre, finely bordered with brown and with faintly defined light shafts. Sickles black, side-hangers black, laced with chamois, with well-defined light shafts. Breast nankin, each feather finely laced with ochre (yellowish-buff), the shafts being distinct and clear. Thighs same colour as breast, abdomen and underparts greyish-brown, with silky, golden barb-shaped tips.

The general effect is that in this variety all the upper parts are dark and the lower parts light, giving the appearance of being covered with a dark chequered cloak. The dominating dark tint is chocolate-black, with a soft silvery lustre, known amongst artists as " umber ". The general light tone is nankin or yellow ochre, and well-defined light shafts are important.

Plumage, Female: Head, face and neck covered with feathers which increase in size as they near the body, ground colour umber with very fine gold lacing. Neck velvety, darker than the back and clearly detached from it. Shaft and lacing clearer and more golden towards the breast. Back covered with umber-coloured feathers having a silvery velvety lustre, each feather dark, finely laced with chamois and with bright nankin shafts showing in strong contrast. Wings the same colour as back, dark umber, finely laced with chamois, feathers broader and brighter towards lower part of wing. Primaries are hidden when wing is closed and are dark intense umber. Tail plumage and cushion similar to back and of same character. Breast, clear even nankin, the shafts pale and distinct, feathers nearing the wings finely and progressively bordered with dark umber, forming a distinctive colour-pattern.

In both sexes: Eyes dark brown (nearly black) with black pupils. Legs and feet slate-grey. Beak and nails horn-coloured.

Defects to be avoided: Salmon or brownish colour on breast.

Plumage, Blue Quail, Male and Female: Similar to the Quail in all respects except that black markings are replaced by blue.

Plumage, Cuckoo, Male and Female: Uniformly cuckoo-coloured, with transverse bars of dark bluish grey on light grey ground. Each feather must have at least three bars.

In both sexes: Eyes orange-red; legs, feet, beak and nails white, often spotted with bluish-grey in young birds.

Defects to be avoided: Feathers white or spotted with white, excessive number of black feathers, red on shoulders, wings and hackle.

Plumage, Black Mottled, Male and Female: All feathers black with green metallic lustre, regularly tipped with white, tips varying in size with the feather. Excessive white markings or uneven distribution to be avoided.

In both sexes: Eyes dark red; legs and feet slate-blue or blackish; beak and nails dark horn.

Plumage, Black, Male and Female: Black all over with metallic green lustre, avoiding false colouring.

In both sexes: Eyes black; legs and feet blue (blackish in young birds); beak and nails black or very dark horn.

Plumage, White, Male and Female: Clear white throughout, avoiding false colours, straw tinge or yellow tint on back.

In both sexes: Eyes orange-red; legs, feet, beak and nails white.

Other Colours: These include Laced Blue (Andalusian type—a diffusion of black and white); Self-blue (true-breeding pale blue); Blue-mottled (similarly marked to Black-mottled); Ermines (black-pointed whites); Fawn Ermines (black-pointed fawns or pale buffs); Partridges; Silvers; Golds and Spangles.

Not all these colours are regularly seen in this country, but there is practically no limit to the sub-varieties capable of being produced in these very charming breeds.

There are also other Bearded Belgian Bantams which are not yet standardised in this country. Of these the most noteworthy is the Barbu de Watermael, which has clean legs, a small " flying " crest or tassel, tri-lobed beard, and neat rose comb with three separate and distinct leaders or spikes.

BRAHMA

Nowadays these are not often seen, but good specimens are some of the most attractive bantams known.

GENERAL CHARACTERISTICS: MALE

Carriage: Upstanding; sedate but active, not as low as the Cochin.

Type: Body broad, deep and square, full breasted, horizontal in keel. Back short, with saddle rising from middle of back until it reaches tail coverts. The body should give an impression, to the eye, of much greater depth than length. Wings medium size, free from twisted feathers or slipped flights, lower line horizontal, with ends tucked under saddle hackle, which should be long and profuse. Tail moderate in length, rising from saddle and carried nearly upright, with well-spread quill feathers and abundant broad coverts, well curved and almost hiding the main tail feathers.

Head: Small, short, of medium breadth, with prominent brows of " Asiatic " character. Beak short and strong. Eyes large and full. Comb triple or " pea ", small, close fitting. Face smooth, free from feathers or hairs. Ear-lobes long and fine, wattles small, fine and rounded, both free from feathers.

Neck: Long, with plentiful hackles reaching well down the shoulders. An indication of breed character is a depression at the back of the neck, between the head feathers and the upper hackle.

Legs and Feet: Legs moderately long, powerful, well apart and plentifully feathered. Thighs large, hidden by lower breast feathers and body fluff, which is very abundant and soft, covering the abdomen and standing out round the thighs. Hocks plentifully covered with soft broad feathers. Profuse shank feathering standing well out from legs and toes, extending to the ends of middle and outer toes. Hock furnishing may be of quill-feather, but vulture hocks are undesirable. Toes, four, straight and spread.

Plumage and Fluff: Profuse, but not as soft as in the Cochin or Pekin.

FEMALE

The neck and legs are rather short; otherwise general characteristics are similar to those of the male, allowing for the natural sexual differences.

COLOUR

Two varieties only are standardised—Dark and Light.

Plumage, Dark Variety, Male: Silver-white head. Neck and saddle hackle silver-white with sharp brilliant black striping free from shaftiness. Back silver-white, but glossy black laced with white between shoulders. Breast and underparts intense glossy black. Wing bows silver, primaries black with white outer edging. Outer web of secondaries partly white, remainder black. Wing coverts glossy black. Tail black, tail coverts edged with white. Footings and leg feather black, or black broken by white.

Plumage, Female: Silver-white head, striped with black or grey.

Neck hackle similar to male, but may be pencilled instead of striped. Tail black, edged with grey, or pencilled. Remainder of plumage any shade of soft clear grey, finely pencilled with black or dark grey; the markings numerous, sharply defined and uniform, concentric in character, following outline of feather.

Plumage, Light Variety, Male: Head and neck as described for Darks. Saddle white, but slight black striping permissible in birds with dense black in neck hackles. Primaries black, or edged with white. Secondaries white outer web, part black inner web. Tail black, but may be edged with white. Remainder clear white top colour. Undercolour may be white, blue or slate. Shank and foot feather black and white.

Plumage, Female: Neck hackle silver-white striped with black, the black centre of each feather entirely surrounded by a white fringe. Otherwise the female colouring is similar to the male.

In both sexes: For both varieties the beak is yellow, or yellow shaded with black. Eyes orange-red. Comb, face, ear-lobes and wattles bright red. Legs and feet orange or yellow.

Suggested Weights: Male 28 to 32 oz. Female 24 to 28 oz.

SCALE OF POINTS

COLOUR AND MARKINGS	40
HEAD POINTS	10
CARRIAGE	10
TYPE AND FEATHER	15
LEGS AND FEET	10
SIZE AND CONDITION	15
	100

Serious Defects: Comb other than pea type. Twisted hackle. Faulty wing feather. Lack of leg feather and footings. Buff, red or yellow in plumage.

COCHIN OR PEKIN

THIS is a genuine bantam breed, very old and probably having no real relationship to the large breed of Cochins. It was imported from Pekin in the middle of the 19th century, hence its name (it is

224

Cochin
White Male,
Mottled Female

still generally called the Pekin bantam). The additional name " Cochin " was adopted in recent years to encourage production of large breed character.

GENERAL CHARACTERISTICS: MALE

Carriage: Bold, rather forward and low, the head very little higher than the tail.

Type: Body short and broad. Back short, increasing in breadth to the saddle, which should be very full, rising well from between the shoulders and furnished with long soft feathers. Breast deep and full. Wings short, tightly tucked up, the ends hidden by saddle hackle. Tail very short and full, soft and without hard quill feathers, with abundant coverts almost hiding main tail feathers, the whole forming one unbroken duplex curve with the back and saddle. General type as much like a ball as possible.

Head: Skull small and fine. Beak rather short, stout, slightly curved. Eyes large and bright. Comb single, small, firm, perfectly straight and erect, well serrated, curved from front to back. Face smooth and fine, ear-lobes smooth and fine, preferably nearly as long as the wattles, which are long, ample, smooth and rounded. (*Note:* A very desirable exhibition character would be the typical Asiatic eye and brow of the large Cochin.)

Neck: Short, carried forward, with abundant long hackle reaching well down the back.

Legs and Feet: Legs short and well apart. Stout thighs hidden by plentiful fluff. Hocks completely covered with soft feathers curling round the joints (stiff feathers forming " vulture hocks " are objectionable but not a disqualification). Shanks short and thick, abundantly covered with soft outstanding feathers. Toes, four, strong and straight, the middle and outer toes plentifully covered with soft feathers to their tips.

Plumage: Very abundant, long and wide, quite soft with very full fluff.

FEMALE

With the exception of the back (rising into a very full and round cushion) the general characteristics are similar to those of the male, allowing for the natural sexual differences.

COLOUR

Plumage, Black Variety, Male and Female: Rich sound black with lustrous beetle-green sheen throughout, free of white or coloured feathers. (*Note:* Some light undercolour in adult males is permissible so long as it does not show through.)

Plumage, Blue Variety, Male and Female: A rich pale blue (pigeon blue preferred) free from lacing, but with rich dark blue hackles, back and tail in the male.

Plumage, Buff Variety, Male and Female: Sound buff, of a perfectly even shade throughout, quite sound to roots of feathers, and free from black, white or bronze feathers. The exact shade of buff is not material so long as it is level throughout and free from shaftiness, mealiness or lacing. (*Note:* A pale " lemon buff " is usually preferred in the show pen.)

Plumage, Cuckoo Variety, Male and Female: Evenly barred with dark slate on light French-grey ground colour.

Plumage, Mottled Variety, Male and Female: Evenly mottled with white at the tip of each feather on a rich black with beetle-green sheen.

Plumage, Partridge Variety, Male: Head dark orange red, neck hackle bright orange or golden red, becoming lighter towards the shoulders and preferably shading off as near lemon colour as possible, each feather distinctly striped down the middle with black, and free from shaftiness, black tipping or black fringe. Saddle hackle to resemble neck hackle as nearly as possible. Breast, thighs, underparts, tail, coverts, wing butts and foot feather, hock feather and fluff lustrous green-black, free from grey, rust or white. Back, shoulder coverts and wing bow rich crimson. Primaries black, free from white or grizzle. Secondaries black inner web, bay outer, showing a distinct wing bay when closed.

Plumage, Female: Head and neck hackle light gold or straw, each feather distinctly striped down middle with black. Remainder clear light partridge brown, finely and evenly pencilled all over with concentric rings of dark shade (preferably glossy green-black). The whole of uniform shade and marking, and the ground colour of the soft brown shade frequently described as the colour of a dead oak leaf, with three concentric rings of pencilling or more over as much of the plumage as possible.

Plumage, White Variety, Male and Female: Pure snow-white, free from cream or yellow tinge, or black splashes or peppering.

In both sexes: Beak yellow, but in dark colours may be shaded with black or horn. Eyes red, orange or yellow---red preferred. Comb, face, wattles and ear-lobes bright red. Legs and feet yellow. (Dark legs are permissible in blacks if the soles of the feet and back of shanks are yellow.)

Suggested Weights: British Bantam Association—Male 24 to 28 oz. Female 20 to 24 oz. Club Standard—Male not to exceed 24 oz. Female 20 oz.

SCALE OF POINTS

COLOUR AND MARKINGS	20
FLUFF AND CUSHION	15
LEG AND FOOT FEATHER	10
SIZE AND WEIGHT	10
TYPE AND CARRIAGE	15
HEAD	10
LENGTH OF SHANK	10
CONDITION	10
	100

Serious Defects: Twisted or drooping comb. Slipped wings. Legs other than yellow (except in blacks). Eyes other than red, orange or yellow. Any deformity.

FAVEROLLE

THE Ermine, Salmon and White varieties of this breed have been bantamised only in recent years; it would not be too unkind to say that they have not been taken up widely by exhibitors. The birds should be true miniatures of the large fowls.

GENERAL CHARACTERISTICS : MALE

Carriage: Active and alert.

Type: Body deep, thick and " cloddy ". Back flat and " square ", i.e., very broad across shoulders and saddle. Breast broad, keel-bone very deep and well forward in front, but not too rounded (a hollow breast is very objectionable). Wings small, prominent in front, carried closely. Tail moderately long, somewhat upright, and with broad feathers (flowing tail, either low or on a level with the back, is very objectionable).

Head: Broad, flat and short, free from crest. Beak short, stout. Eyes prominent. Comb single, medium size, upright, with four to six serrations, smooth and free from coarseness or any side work.

228

Face muffled; muffling full, wide, short and solid. Ear-lobes and wattles small, of fine texture, and partly concealed by the muffles.

Neck: Short and thick, especially near the body, which it should be well let into.

Legs and Feet: Legs short and stout. Thighs wide apart. Shanks straight and medium length with width between them, and sparsely feathered to the outer toe. Narrowness or any tendency to in-kneed is very objectionable. Toes, five, the front three long, straight and well spread, the outer toe sparsely feathered, the fourth toe (quite divided from the fifth) on the ground and well back, the fifth turned up the leg.

FEMALE

The general characteristics are similar to those of the male, allowing for the natural sexual differences, with the following exceptions. Comb much smaller in proportion, back longer in proportion, neck straighter, keel bone longer and deeper and the tail carried midway between upright and drooping.

COLOUR

Plumage, Ermine Variety, Male and Female: Head and neck plumage white striped with black, the centre of each feather entirely surrounded by a white margin. Wings white with black in flights. Tail black. Remainder pure white.

In both sexes: Beak horn or white. Eyes grey or hazel, comb red. Face, wattles and ear-lobes red, partly concealed by muffling. Legs and feet white.

Plumage, Salmon Variety, Male: Beard and muff black. Hackles straw. Back, shoulders and wing bows bright cherry mahogany. Breast, thighs, underfluff, tail and shank feathering black. Wing bar black; primaries black; secondaries white outer edge, black inner edge and at tips.

Plumage, Female: Beard and muff creamy-white. Breast, thighs and fluff cream. Remainder wheaten brown; head and neck striped with dark shade of the same colour (free from black) and wings softer and lighter than back. Primaries, secondaries and tail wheaten brown.

In both sexes: Beak, etc., as in the Ermine.

Plumage, White Variety, Male and Female: Pure white. In both sexes. Beak, etc., as in the Ermine.

Standard Weights: Male $2\frac{1}{2}$ to 3 lb. Female 2 to $2\frac{1}{2}$ lb.

SCALE OF POINTS

UTILITY QUALITIES, SIZE AND CONDITION	20
TYPE	25
COLOUR	20
BEARD AND MUFFLING	20
FORMATION OF FEET AND TOES	5
FOOT FEATHER	5
COMB	5
	100

Serious Defects: Side sprigs in single comb. Badly lopped comb. Wry or squirrel tail. White in ear-lobe. Crooked back or breast. Deformed beak. Little or no beard, mufflings, and foot feathers.

Disqualification: Other than five toes on each foot.

FRIZZLE

THE Poultry Club standardises Frizzles in this country as a recognised breed, but on the Continent the word " frizzle " merely denotes a feather character, and specimens are usually found in other breeds, especially Japanese.

GENERAL CHARACTERISTICS: MALE

Carriage: Erect, active and strutting.

Type: Back broad and short. Breast round and full. Wings long and drooping. Tail full and loose, carried high. Sickles full, side hangers plentiful. (*Note:* Lyre-shaped main sickles in males have been favoured in the past but are not standardised.)

Head: Fine skull. Beak short and strong. Eyes full and bright. Comb single, medium-sized and upright. Face smooth. Ear-lobes and wattles moderate in size.

Neck: Moderate length with abundantly curled hackles showing a decided " mane ".

White Frizzle
Male

Blue Frizzle
Female

Legs: Moderately short, set well apart, with shanks quite free from feathers. Toes, four, well spread and fine.

Plumage: Fairly long and crisp, each feather broad and curled back towards the head; the curl as close, even and abundant as possible. Hackle full and well curled.

FEMALE

The general characteristics are similar to those of the male, allowing for the natural sexual differences, except that the comb is much smaller in proportion and the neck not so abundantly frizzled (i.e., the " mane " is smaller, due to less hackle).

COLOUR

Plumage, Male and Female: The plumage is of pure even colour throughout in self-coloured varieties. Colour of Columbians as in Wyandottes. Other " natural " colours as in Old English Game.

  In both sexes: Beak yellow or horn. Legs and feet yellow in Buffs, Reds, Columbians, Whites, Duckwings, Black-reds, Spangles or Mottles, and Cuckoos; and black, blue or willow in dark varieties. (There are variations in leg colour, and yellow legs are frequently demanded in Blacks, though not so standardised.) White legs are permissible where so standardised in O.E. Game colours. Eyes brilliant red. Comb, face, wattles and ear-lobes bright red.

Suggested Weights: Male 24 to 28 oz. Female 20 to 24 oz.

SCALE OF POINTS

HEAD AND COMB	5
LEGS AND FEET	5
PLUMAGE COLOUR	15
SIZE	10
CURL	25
FEATHER QUALITY	20
TYPE AND SYMMETRY	10
CONDITION	10
	100

Serious Defects: Want of curl. Other than single comb. Drooping comb. White lobes. Narrow or soft feather. Long tail. Any deformity.

232

HAMBURGH

These exist in two varieties, each of two colours—Spangled and Pencilled in Silvers and Golds. There are no Black or White varieties whose places are filled by Rosecombs of those colours.

GENERAL CHARACTERISTICS: MALE

Carriage: Graceful, alert and bold.

Type: Body moderately long but compact, fairly wide at shoulders. Back flat, tapering to tail. Breast well rounded. Wings large and neatly tucked. Tail long and sweeping, carried well up, sickles broad and furnishings very plentiful.

Head: Fine. Beak short and well curved. Eyes bold and full. Comb rose, of medium size, square-fronted, firmly set, tapering steadily to a long fine-ended leader, straight and level, without downward tendency. Top of comb without hollows and full of fine work. Face smooth and hairless. Ear-lobes smooth, round and flat, not dished. Wattles smooth, round and fine.

Neck: Medium length, plentifully furnished with long hackles well covering the shoulders.

Legs and Feet: Legs of medium length. Thighs slender, shanks fine and round, free from feathers. Toes, four, slender and well spread.

FEMALE

The general characteristics are similar to those of the male, allowing for the natural sexual differences.

COLOUR

Plumage, Gold Pencilled Variety, Male: Bright red bay or golden chestnut, except the tail, which is black, the sickle feathers and coverts being laced all round with a narrow gold edging.

Plumage, Female: Ground colour similar to the male. Except on neck hackle (which should be as clear of markings as possible) each feather to be distinctly and evenly " pencilled " or barred straight across with fine parallel lines of rich green-black. Barring and ground-colour to be the same width, and the finer and more numerous the markings the better.

Plumage, Silver Pencilled Variety, Male and Female: Similar to Golds except that ground colour and lacing are silvery-white.

Plumage, Gold Spangled Variety, Male: Ground colour rich bay or mahogany. All markings, tipping, spangling and tail rich green-black. Hackles and back striped with black down the centre of each feather. Dagger-shaped tips to wing bows. Wing bars two distinct rows of large round black spangles, running parallel across each wing with gentle curve. Secondaries tipped with large round spangles forming the steppings. Breast and underparts tipped throughout with round green-black spots or spangles, small at throat and increasing in size towards the thighs but not large enough to overlap.

Plumage, Female: Similar generally to the male. Tail coverts black with sharp lacing or edging of gold. Remaining feather tipped throughout with green-black spangles as large and round as possible without overlapping. Steppings as in the male. Spangling to commence high up the throat.

Plumage, Silver Spangled Variety, Male: Ground colour pure silver, spangling and tipping rich green-black. Hackles, shoulders and back marked on each feather with small dagger-like tips. Wing bows with dagger-shaped tips, increasing in size until they merge in what is known as the third bar. Bars and secondaries, breast and underparts as in the gold spangled. Tail feathers ending with bold half-moon shaped spangles; sickles with large round spangles on each feather. Tail coverts similar but spangling not so big.

Plumage, Female: Ground colour and spangling as described for male. Hackle marked from the head with dagger-shaped tips, increasing in width until they merge into spangles at bottom. Wing bars and secondaries as in the male. Each tail feather to have half-moon spangle at end. Tail coverts reaching half way up the true tail and forming a row of round spangles across each side. Remainder marked as gold female.

In both sexes: Beak dark horn; eyes, comb, face and wattles red; ear-lobes white; legs lead-blue.

Suggested Weights: Male 24 to 28 oz (max.). Female 22 to 26 oz (max.). (The Pencilled varieties are usually considerably larger.)

SCALE OF POINTS

MARKINGS.............................. 60
HEAD, COMB, FACE AND LOBES........... 20
COLOUR................................ 10
TYPE, STYLE AND CONDITION............ 10
                                      ———
                                      100
                                      ═══

(In Pencilled males, the points allotted to markings
are divided between tail and colour.)

Serious Defects: Red ear-lobes. White in face. Squirrel or wry tail, or any other deformity.

HOUDAN________________________________

Houdans are one of the newest breeds of bantams accepted to Standards, and they are as yet in very few hands.

GENERAL CHARACTERISTICS: MALE

Carriage: Bold and lively.

Type: Body broad and lengthy, as in the Dorking. Tail full; sickles long and well curved. Wings medium, carried close.

Head: Large, with a decided protuberance on top, and a full and compact crest, round on top and not divided or split, composed of feathers similar to hackle feathers, set so as to expose the comb and not obstruct sight. Beak short and stout, well curved, with wide nostrils. Eyes bold. Comb leaf (somewhat resembling a butterfly) fairly small, and both sides equal. Face muffed and bearded. Muffling large, full and compact, fitting round to back of eyes and almost hiding the face. Ear-lobes small and completely hidden under muffling. Wattles small, well rounded and almost hidden by the beard.

Neck: Medium length, with abundant hackles reaching well down the back.

Legs and Feet: Legs short and stout, free of feathers and well apart. Toes, five, these and body shape suggesting part-Dorking ancestry.

FEMALE

With the exception of the crest (which must be globular) the

235

characteristics are generally similar to those of the male, allowing for the natural sexual differences.

COLOUR

Plumage, Male and Female: Glossy green-black ground with pure white mottling evenly distributed all over, except on flights and secondaries. Sickles and tail coverts in the male irregularly edged with white. In young birds black usually predominates and mottling becomes gayer with age.

In both sexes: Beak horn, eyes red. Comb, face and wattles bright red. Ear-lobes white or tinged with pink. Legs and feet white, mottled with black or lead-blue.

Suggested Weights: Male 24 to 28 oz. Female 22 to 26 oz.

SCALE OF POINTS

	Male	Female
TYPE	12	10
COLOUR	15	15
MUFFLING	8	12
SIZE	18	20
COMB	15	8
CREST	12	15
LEGS AND FEET	10	10
CONDITION	10	10
	100	100

Serious Defects: Red or straw-coloured feathers. Spur outside the shank. Feathers on shanks or toes. Loose crest obstructing sight. Other than five toes on each foot. Any deformity.

INDIAN AND JUBILEE INDIAN GAME

THESE miniatures are well established. They are not recognised as Game birds in spite of their name; and although they may compete in classes specified for " hard feather " they may not compete in classes listed for Game birds. Where no " hard feather " or breed classes are scheduled, they must compete in the A.O.V.

Indian Game
Male and Female

section.   Jubilee are similar to the Indian except that where Indians are black, Jubilees are white.   Both colours are frequently interbred.

GENERAL CHARACTERISTICS: MALE

Carriage: Powerful and upright, commanding and courageous, active and vigorous.

Type: Body very thick and compact, very broad at shoulders and with prominent wing butts.   Back short and flat.   Body must not be flat-sided or hollow-backed.   Breast wide, prominent and well rounded.   Body tapers towards tail.   Wings short and muscular, high in front, carried close, and round at ends.   Tail of medium length, with a slight droop; sickles and coverts narrow, short, close and hard.

Head: Broad, not so keen as in Game nor as thick as in the Malay. Beak stout, short and well curved, giving the head a powerful appearance.   Comb pea, small and close fitting.   Eyes full and bold, slightly beetle-browed, but not showing such a cruel expression as the Malay.   Face smooth and fine, ear-lobes and wattles small. Throat not so bare as in O.E.G., but dotted with fine feathers. (*Note:* Males are usually dubbed, leaving skull and lower jaws smooth and free from ridges.)

Neck: Medium length and slightly arched, hackle feathers short and just covering base of neck.

Legs and Feet: Legs very strong and thick.   Thighs round and stout.   Shanks short, well scaled, wide apart and free of feathers. Toes, four, strong, straight and spread, the rear toe low and nearly flat on ground.

Handling: Flesh firm.

Plumage: Short, close and hard.

FEMALE

The general characteristics are similar to those of the male, allowing for the natural sexual differences.   The tail is closely folded and carried somewhat higher than in males.

COLOUR

Plumage, Male: Head, neck, breast, thighs, underfluff and tail black with rich green sheen, the base of neck and tail hackles broken slightly with bay or chestnut.   Wing bows and shoulders glossy green-black, broken with bay or chestnut.   Back, tail coverts and furnishings all green-black touched with chestnut or bay.   Wing ends when closed show a triangle of bay or chestnut.   Primaries and secondaries glossy green-black, marked with chestnut, the primaries having a narrow chestnut lacing on outer web.

Plumage, Female: Ground colour chestnut, nut-brown or mahogany-brown. Head, hackle and throat glossy green-black. Hackle at base of neck beetle-green with bay or chestnut centres. The whole of the rest of the body so far as possible to be double-laced with rich glossy green-black markings, the inner lacing being particularly distinct. Belly, thighs and underparts similarly marked but less distinct. Markings on wing bows and shoulders are clearest of all. Tail coverts have the same type of marking, but seldom so distinct. Primary and secondary feathers black, peppered or marked with chestnut. Black markings should look as if embossed or raised; and in a good specimen there is often a black-marked centre to feathers within the double-lacing.

In both sexes: Beak horn or yellow, or both; eyes vary from pearl or pale yellow to pale red. Face, comb, lobes and wattles bright red. Legs and feet rich yellow or orange, the deeper the better.

Suggested Weights: British Bantam Association—Male 36 to 40 oz. Female 32 to 36 oz. Club Standard—Male 40 to 48 oz. Female 32 to 40 oz. The Indian Game Club does not issue definite weight standards for bantams. The Club standard weights detailed above are suggestions only. (These weights are often exceeded. Excess size should be penalised.)

JUBILEE INDIAN GAME

GENERAL CHARACTERISTICS: As for Indian Game.

COLOUR

Plumage, Male: Head, neck, breast, underfluff, thighs and tail white. Clear breast is desirable. Shoulders and wing bows white, slightly broken with bay or chestnut in centre of feather. Tail coverts white; back white, touched on ends of feathers with bay or chestnut. Primaries and secondaries white with bay markings, showing a triangular patch of bay or chestnut when wing is closed.

Plumage, Female: Ground colour chestnut, nut-brown or mahogany. Head, hackle and throat white. Primaries white, peppered or marked with chestnut on inner web. Secondaries white on inner web, outer web chestnut, delicately edged with white. Main tail feathers white. Remainder of plumage one even ground colour, double-laced with white; the inner or second lacing being most distinct. Tail coverts are usually not so distinctly marked, and underparts and thighs may run off into indistinct lacing.   Wing coverts may be somewhat peppered.

In both sexes: Beak, eyes, head points and legs generally as described for Indian Game.

In all other respects the Indian Game standard should be followed.

SCALE OF POINTS

The Indian Game Club does not publish a scale of points, but the following is generally accepted:

SKULL, EYES AND BROWS	9
BEAK, WATTLES, LOBES AND COMB	8
NECK	3
SIZE	10
CONDITION	8
BODY AND THIGHS	10
BACK AND BREAST	16
WINGS, TAIL AND LEGS	24
CARRIAGE	12
	100

Serious Defects: Crooked back.   Knock-knees or bow-legs.   Wry or squirrel tail.   Single or walnut comb.   White in hackle.   Red or twisted hackles.   Soft feather.   Flat shins.   Crooked breast or toes.   Mealy ground colour in hens.

JAPANESE

TRUE bantams of great antiquity, these are without counterparts in large breeds.   They are the shortest-legged of all varieties.   The standard here given is that of the International Japanese Club, which was approved and adopted by the British Club in 1937.

GENERAL CHARACTERISTICS: MALE

Carriage and Appearance: Very small, low built, broad and cobby with deep full breast and full-feathered upright tail.   Appearance somewhat quaint due to a very large comb, dwarfish character and waddling gait.   Plumage very full and abundant.

Japanese
Black Male, Buff
Female

Type: Back very short, wide, and seen from the side it forms the shape of a small letter U, the sides being formed by the neck and tail. This shape, however, is almost lost in fully feathered males. Saddle hackles rich and long. Body short, deep and broad. Breast very full, round and carried prominently forward. Wings long with the tips of the secondaries touching the ground immediately under the end of the body. Thighs very short and not visible. Tail very large and upright. The main tail feathers should rise above the level of the head about one third of their length, spreading well and with long sword-shaped main sickles and numerous soft side hangers. The tail may touch the comb with its front feathers, but must not be set so as to lean forward at too sharp an angle.

Head: Large and broad, beak strong and well curved, eyes large. Comb single, large (the larger the better), coarse grained, erect and evenly serrated with four or five points. The blade of the comb should follow the nape of the neck. Face smooth, ear-lobes medium size, red and free from all traces of white. Wattles pendent and large.

Neck: Rather short, curving backwards and with abundant hackle feathers which should well drape the shoulders.

Legs and Feet: Shanks very short, clean (free from feather), strong and sharply angled at the joints. The shanks to be so short as to be almost invisible. Toes four, straight and well spread.

Weights: Male 18 to 22 oz. Female 14 to 18 oz.

FEMALE

The general characteristics should follow closely those described for the male regarding type. Breast should be all as described for the male. Tail well spread and rising well above the head. The main tail feathers broad, the foremost pair being slightly curved (sword shaped). Comb large, evenly serrated and preferably erect, although falling to one side being no defect.

COLOUR

Black-tailed Whites: Body feathers white, wing primaries and secondaries should have white outer and black inner webs, the closed wings look almost white. Main tail feathers black or black with white lacing. Main sickles and side hangers black with white edging. Eyes red. Legs yellow.

Black-tailed Buffs. The same markings as Black-tailed Whites, except that the white is replaced by buff.

242

Buff Columbian: Rich even buff, wing primaries and secondaries buff with black inner webs, the closed wings look almost buff. Sickles and side hangers black with buff edges. Neck hackle feathers buff with black centre down each, the hackle to be free from black edges. Eyes red. Legs yellow.

White: Pure white without sappiness. Eyes red. Legs yellow.

Black: With red comb and face. Deep full black with a green sheen. Legs yellow, black permitted on shanks but underside of feet must be yellow. Eyes black.

Greys: Birchen Grey. The hackle, back, shoulder coverts and wing bows silver-white. The neck, saddle hackle and saddle with narrow black striping. Remainder charcoal-grey or black, the breast feathers having a narrow light grey or silver edging around each feather, thereby giving a laced appearance. In females the neck hackle light grey or silver with narrow black striping. Breast laced as for males, remainder charcoal-grey, or black. Birchen Grey may have dark comb, flesh and face.
Dark Grey: As above but breast charcoal-grey or black. Female almost all dark with striped neck hackle.
Light Grey: As Birchen Grey but breast silver-grey, mealy in both sexes.
All Greys to have yellow or willow legs with yellow soles. Beak shaded with black or horn. Eyes red, orange or dark brown iris.

Mottleds: Black Mottled. All feathers should be black with white tips. The amount of white may vary, but the ideal is between $\frac{3}{8}$ in and $\frac{1}{2}$ in. Tail and wings similar but more white permitted.
Blue Mottled. These are as above but blue instead of black.
Red Mottled. These are as Black Mottled but red instead of black. In all mottleds the legs should be yellow or willow. Beaks to match legs. Eyes red or orange.

Blue: Self Blue. All feathers blue, neck feathers may be a darker blue than the remainder. Legs slate or willow. Eyes orange.
Lavender Blue. All feathers a lavender blue to skin, even shade throughout. Legs slate or blue. Eyes orange or dark brown.

Cuckoo: The feathers throughout including body, wing and tail to be generally uniformly cuckoo coloured with transverse bars of dark bluish-grey on a light grey ground. Legs yellow. Beak yellow marked with black. Eyes orange or red.

Red: All feathers deep red, solid to skin an even shade throughout. Legs and beak yellow. Both legs and beak may be marked with red. Eyes red.

Tri-coloured: The colours white, black and brown or dark ochre should be as equally divided as possible on each feather. Legs yellow or willow. Eyes red or orange.

Black-red: Wheaten Bred. Partridge Bred.

Brown-red.

Blue-red.

Silver Duckwing.

Golden Duckwing.

The latter four colours all as described for Old English Game.

The following secondary colours permitted: Ginger, Blue-dun, Honey-dun, Golden Hackled, Furnace.

SCALE OF POINTS

TYPE	55
SIZE	15
CONDITION	15
COLOUR	10
LEG COLOUR	5
	100

Serious Defects: Narrow build. Long legs. Long back. Wry tail. Tail carried low. Deformed comb and lopped comb on males. High wing carriage. White in lobes. Any physcial deformity.

LEGHORN

Leghorn bantams are exhibited in many colours and resemble their larger counterparts to a remarkable degree. Most popular are Whites and Browns.

GENERAL CHARACTERISTICS: MALE

Carriage: Very sprightly and alert, completely without stiltiness.

Type: Body wedge shaped, wide at shoulders, narrowing slightly to root of tail. Back long, sloping slightly to tail. Breast round and prominent. Wings large, carried close and well tucked up. Tail moderately full, carried at an angle of 45 degrees from the back.

Head: Fine. Beak stout, with point clear of front of comb. Eyes prominent. Comb single or rose. (Rose-combs are seldom seen.) The single perfectly straight and erect, large but not overgrown, deeply and evenly serrated, the spikes broad at base. Comb to extend well beyond back of head and to follow, without touching, the line of head; without thumb marks or side sprigs. The rose to be moderately large, firm, not so overgrown as to obstruct the sight, the leader extending straight back and not following line of head. Top of comb covered with small close spikes of even height, and free from hollows. Face smooth. Ear-lobes well developed and rather pendent, matched, smooth, open and free from folds. Wattles long and thin.

Neck: Long, profusely covered with hackle feathers.

Legs and Feet: Legs moderately long. Shanks fine and round, free of feathers (flat shins objectionable). Toes, four, long, straight and well spread.

Plumage: Silky in texture, free from excessive feather or woolliness.

Handling: Firm, with plenty of muscle.

FEMALE

In the single-combed variety the comb falls gracefully over either side of the face without obstructing the sight. (Erect combs in adult females are objectionable.) The tail is carried closely and not at such a high angle. Otherwise the general characteristics are similar to those of the male, allowing for the natural sexual differences.

COLOUR

Plumage, Black Variety, Male and Female: Black with rich green sheen, perfectly free of any other colour. Undercolour as dark as possible.

Plumage, Blue Variety, Male and Female: Level shade of blue throughout, without lacing, permitting darker shade in male hackles and neck of female.

Plumage, Brown Variety, Male: Neck hackle rich orange-red, shading off to lemon colour at tips, striped with black, crimson-red at front below wattles. Back, wing bow and shoulders crimson-red or maroon. Wing coverts steel-blue with green sheen. Primaries brown; secondaries deep bay on outer web, black on inner. Saddle rich orange-red, shading off to match neck hackle as nearly as possible. Breast and underparts glossy black, completely free from brown. Tail black with rich green sheen; any white in tail very objectionable. Tail coverts black edged with brown.

Female: Neck hackle rich golden-yellow, broadly striped with black. Breast salmon-red, running into maroon round head and wattles and shading into ash-grey at thighs. Body colour soft rich brown, very closely and evenly pencilled with black, free from light shafts to feathers, and wings free from red or rust. Tail black, outer feathers pencilled with brown.

Plumage, White Variety, Male and Female: Pure white throughout, free from straw tinge or other colour.

In both sexes: Beak yellow or horn. Eyes red. Comb, face and wattles bright red. Ear-lobes pure opaque white, kid-like in texture, or cream—white preferred. Legs and feet yellow or orange.

Other colours are occasionally seen but are not standardised in bantams.

Standard Weights: Male 32 oz. (max.). Female 28 oz. (max.).

SCALE OF POINTS

	Whites	Blacks	Browns	Blues
COMB	10	10	12	10
LOBES	10	10	11	10
EYES	5	5	5	5
LEGS	10	10	—	10
BREAST BONE	5	—	—	—
CONDITION	10	10	12	10
TYPE	20	15	15	15
SIZE	15	15	15	10
PLUMAGE AND COLOUR	15	25	30	30
	100	100	100	100

Serious Defects: Comb lopped in males, or erect in females. Rose-comb obstructing the sight. Ear-lobes red. White in face. Legs other than yellow or orange. Wry or squirrel tail. Any deformity. In Blacks, dark legs or eyes. In Browns, white in feather.

MALAY

THIS is an old-established bantam breed, not nowadays popular except in Cornwall and Devon. They are large in comparison with other bantams; and it is not easy to reduce them further without losing the typical large breed character.

GENERAL CHARACTERISTICS: MALE

Carriage: Gaunt, fierce and very erect, high in front, drooping at stern and straight at hocks, with hard, clean, cut-up appearance at stern.

Type: Body short, wide in front and tapering, with broad square shoulders and prominent wing butts carried high and devoid of feathers on points. Back short and sloping with convex-curved outline and narrow drooping saddle. Breast deep and full, devoid of feathers at point of keel. Wings large and strong, carried close and high. Tail moderate in length, drooping but not whipped, the narrow sickles only slightly curved. The outline of neck, back and upper tail should show three typical curves at nearly equal angles.

Head: Very broad with projecting beetle-brows and a morose, cruel expression. Beak strong, short and well hooked (skull and beak in profile resemble the arc of a circle). Eyes deep set. Comb half-walnut type, small, free from irregularities and set well forward on the head. Face smooth. Ear-lobes and wattles small.

Neck: Long and upright, curved in profile, thick through from gullet to back of skull, throat and gullet bare of feathering. Hackles full at skull but scanty down the neck.

Legs and Feet: Legs long and massive. Thighs muscular, sparsely feathered and with exposed hocks clear of plumage. Shanks clean, well scaled, flat at hocks and rounding to the spurs, which should curve downwards. Toes, four, long and straight, powerful and with strong nails, the hind toe close to ground.

Plumage: Very scanty and short, hard and narrow.

Handling: Very firm-fleshed and muscular.

FEMALE

The general characteristics are similar to those of the male, allowing for the natural sexual differences, but the tail, carried slightly above the horizontal, is short and neither fan-shaped nor whipped.

COLOUR

Several colour-varieties are recognised, among them Black-reds (in which females may be cinnamon, partridge, wheaten or clay in colouring), Whites, Spangles, Blacks, Mottles (black and white), Piles, Duckwings (silver and gold) and Reds. Colours and markings follow the usual Game patterns and descriptions.

In both sexes: Beak yellow to horn. Comb, face, throat, wattles and ear-lobes bright red. Eyes pearl, daw or yellow. Red or foxy eyes are objectionable. Legs and feet rich yellow, with a little duskiness permissible in dark colours.

Desirable Maximum Weights: Male 42 to 48 oz. Female 36 to 40 oz. Most specimens exceed these weights.

SCALE OF POINTS

CURVES, TYPE AND CARRIAGE	20
REACH AND STILTINESS	10
HEAD	15
SHORT, NARROW AND HARD FEATHER	10
CONDITION AND VIGOUR	10
SIZE	15
COLOUR	5
TAIL	5
LEGS	10
	100

Serious Defects: Over-size. Evidence of alien cross. Single or pea-comb. Knock-knees, bow-legs or bad feet. Any deformity.

MARANS

THESE are miniatures of the large breed, and are sponsored by the large breed club.

GENERAL CHARACTERISTICS: MALE

Carriage: Active, compact and graceful.

Type: Body of medium length, with good depth and width throughout; front broad, full and deep. Breast long, well fleshed, of good width, and without keeliness. Tail well carried, high.

Head: Refined. Beak deep and of medium size. Eyes large and prominent; pupil large and well defined. Comb single, medium size, straight and erect, with five to seven serrations, and of fine texture. Face smooth. Wattles of medium size and fine texture.

Neck: Of medium length and not too profusely feathered.

Legs and Feet: Legs of medium length, wide apart, and good quality bone. Thighs well fleshed, but not too heavy in bone. Shanks clean and unfeathered. Toes, four, well spread and straight.

Plumage: Fairly tight and of silky texture generally.

Handling: Firm, as befits a table breed. Flesh white and skin of fine texture.

FEMALE

General characteristics of the female are similar to those of the male, allowing for the natural sexual differences.

COLOUR

Plumage, Dark Cuckoo Variety, Male and Female: Cuckoo throughout, each feather barred across with bands of blue-black. A lighter shaded neck in both sexes and on the back of the male is permissible if definitely barred. Cuckoo throughout is the ideal, as even as possible.

Plumage, Golden Cuckoo Variety, Male: Hackles, bluish-grey with golden and black bars. Neck paler than saddle. Breast bluish-grey with black bars, pale golden shading on upper part. Thighs and fluff light bluish-grey with medium black barring. Back, shoulders and wing bows bluish-grey with rich bright golden and black bars. Wing bars bluish-grey with black bars; golden fringe permissible. Wings, primaries dark blue-grey, lightly barred; secondaries dark blue-grey, lightly barred with slight golden fringe. Tail dark blue-grey barred with black. Tail coverts blue-grey barred with black. General Cuckoo markings.

Plumage, Female: Neck hackles medium bluish-grey with golden and black bars. Breast dark bluish-grey with black bars, pale golden shading on upper parts. Remainder dark bluish-grey with black bars. Cuckoo markings.

Plumage, Silver Cuckoo Variety, Male: Mainly white in neck and showing white on upper part of breast and also on top. Remainder barred throughout, with lighter ground colour than the Dark Cuckoo.

Plumage, Female: Mainly white in neck and, showing white on upper part of breast. Remainder barred throughout with lighter ground colour than in the Dark Cuckoo.

In both sexes: Beak, white or horn. Eyes, red or bright orange preferred. Comb, face, wattles and ear-lobes, red. Legs and feet white.

Weights: Cock 32 oz; Cockerel 28 oz. Hen 28 oz; Pullet 24 oz.

SCALE OF POINTS

TYPE, CARRIAGE AND MERITS (TO INCLUDE
TYPE OF BREAST AND FLESHING, ALSO
QUALITY OF FLESH).................... 40
SIZE AND QUALITY....................... 20
COLOUR AND MARKINGS................. 15
HEAD POINTS 10
CONDITION............................. 10
LEGS AND FEET......................... 5
————
100

Serious Defects: Feathered shanks. General coarseness. Lack of activity. Superfine bone. Any points against utility or reproductive values. **Defects** (for which a bird may be passed): Deformities, crooked breast bone, other than four toes, etc.

MINORCA

THESE should be true miniatures of the large breed. There is at present no separate club, but bantams are catered for by the large breed club.

GENERAL CHARACTERISTICS: MALE

Carriage: Horizontal, but upright and graceful.

Type: Body broad at shoulder, square and compact but fairly long in back, with deep straight keel. Breast full and round. Wings moderate in length, close-fitting. Tail full, with broad long sickles well arched and carried well back.

Head: Long and broad enough to provide firm foundation for a large comb. Beak stout and fairly long. Eyes full, bright and expressive. Comb large, single, arched and evenly serrated, perfectly upright, set firmly on the head, not extending over point of beak, free from twists, side sprigs or thumb marks, reaching well to back of head, and following without touching the line of neck; moderately rough in texture. Five serrations preferred. Face smooth, fine and unwrinkled, free of hair or feather and showing no white. Ear-lobes medium size, almond-shaped, smooth and flat, more elongated than round, widest at top, kid-like in texture, close-fitting, not extending over face and not slack or dished. (*Note:* No definite size of lobe is fixed by standard.) Wattles long and rounded at ends.

Neck: Long, nicely arched, with flowing hackle.

Legs and Feet: Legs and thighs medium length. Shanks strong and free of feathers, straight and wide apart, without stiltiness or knock-knees. Toes, four, fine and well spread.

FEMALE

The general characteristics are similar to those of the male, allowing for the natural sexual differences, but the comb should fall gracefully over one side of the face, in such a way as not to obstruct the sight. (It should be noted that this is a standard requirement, and upright combs in matured females should not be encouraged.)

A rose-combed variety exists in large breeds but not in bantams.

COLOUR

Plumage, Black Variety, Male and Female: Black with brilliant green sheen.

In both sexes: Beak dark horn. Eyes dark. Comb, face and wattles blood red. Ear-lobes pure white. Legs and feet black, or very dark slate in adults.

Plumage, White Variety, Male and Female: Pure lustrous white.

In both sexes: Beak white. Eyes red. Comb, face and wattles blood red. Legs pinky white.

A blue variety is now standardised in large breeds, and blue bantams have been shown at Olympia since the war.

Club Standard Weights: Male 34 oz (max.). Female 30 oz (max.). Smaller specimens to be favoured, other points being equal.

SCALE OF POINTS

FACE	15
LOBES	10
LEGS, EYES AND BEAK	8
SIZE	15
CONDITION	10
COMB	15
COLOUR	10
TYPE AND SHAPE	10
BREAST BONE	7
	100

Serious Defects: White or blue in face. Blushed lobe. Wry or squirrel tail. Feathers on shanks. Foul colour in plumage. Incorrect leg colour. Other than four toes. Crooked breast bone. Any other deformity. Side sprigs in comb.

MODERN GAME

FINE body, " reachiness " and colour are main points in these. The breed is the favourite of " die-hard " showmen.

GENERAL CHARACTERISTICS: MALE

Carriage: Active and upstanding. In the show pen the bird should show plenty of " lift ", as if reaching to its fullest height.

Type: Body short and fine. Back flat, comparatively wide at front and tapering to tail like a flat-iron. Shoulders prominent and carried high. Wings strong and short. Tail short, fine, closely whipped, and carried slightly higher than body-line; sickles narrow, well pointed and only slightly curved.

Head: Long, snaky and narrow between prominent eyes. Beak long, gracefully curved, strong at base. Comb single, small, erect, fine in texture and evenly serrated. Face smooth. Ear-lobes and wattles fine and small. (It is customary to dub Game cocks by removing comb, wattles and lobes. In Modern Game the dubbing is very close and smooth.)

Neck: Long and slightly arched, with hard wiry feathers; feathering not plentiful, particularly at junction of neck and body.

252

Modern Game
White Male,
Black-red Female

Legs and Feet: Legs long and well rounded. Thighs muscular, shanks clean. Toes, four, long, fine and straight, the rear toe flat and straight out, with no tendency to duck-foot.

Plumage: Very short and hard.

FEMALE

The general characteristics are similar to those of the male, allowing for the natural sexual differences.

COLOUR

Colour is very important in Moderns. Varieties include Black-reds, Duckwings, Piles, Brown-reds, Birchens, Blacks, Whites and Blues. Self-colours had been brought to great excellence a few years ago, but are now seldom seen.

Legs and beaks vary with the colour-varieties, from yellow in Piles and Whites through willow in Black-reds to black in Birchens. Eyes similarly vary from bright red to black. Combs and faces vary from bright red to dark purple and black. Leg colours are definitely " tied " to each variety. Thus Whites and Piles must always have yellow legs, while shanks are willow in Duckwings and black in Brown-reds.

Colours are not so numerous as in Old English Game.

Plumage, Black-red Variety, Male: Cap orange-red. Neck hackle light orange, free from black striping. Back and saddle rich crimson. Wing bow orange; wing bar green-black; secondaries, bay outer edge, black inner edge and tips—only the bay showing when wing is closed; primaries black. Remainder green-black.

Plumage, Female: Hackle gold, slightly striped with black, but clear gold on cap. Breast rich salmon, shading off to ash on thighs. Main tail black, but top feathers matching body colour which should have a light partridge-brown ground very finely pencilled with black, with a slight golden tinge over the whole. Body colour should be even throughout, free from any trace of ruddiness (particularly on wings) and with no trace of pencilling on flights.

In both sexes: Beak dark green. Eyes, comb, face, ear-lobes and wattles bright red. Legs willow.

Plumage, Pile Variety, Male: Hackles bright orange-yellow of even shade; dark or washy hackles to be avoided. Back and saddle rich maroon. Wing bow maroon; wing bar white, free from splashes; primaries white; secondaries, dark chestnut outer web,

254

Modern Game
Duckwing Male,
Pile Female

white inner web and tips, only the chestnut showing when wing is closed.   Remainder pure white.

Plumage, Female: Hackle white tinged with gold.   Breast rich salmon-red.   Remainder pure white.

In both sexes: Beak yellow.   Eyes bright cherry-red.   Comb, wattles, ear-lobes and face red.   Legs and feet rich orange-yellow.

Plumage, Brown-red Variety, Male: Hackles, back and wing bow bright lemon;   neck hackle striped down centres with green-black (not brown).   Remainder green-black, the breast feathers delicately edged with pale lemon;   this lacing not extending lower than top of thighs.

Plumage, Female: Neck hackle light lemon colour to top of head, the lower feathers striped with green-black.   Remainder green-black, the breast delicately laced as in the male;   shoulders free from ticking and back free from lacing.

In both sexes: Beak very dark horn or black—the latter preferred. Eyes, comb, face, ear-lobes, wattles, legs and feet black.

Note: There should be only two colours in Brown-reds—lemon and black.   In the male the lemon should be very rich and bright; in the female light.   In both sexes the black should have a bright beetle-green gloss.

Plumage, Birchen Variety, Male and Female: Similar to the Brown-red but silver where the Brown-red is lemon.

Plumage, Golden Duckwing Variety, Male: Hackle creamy white, free from striping.   Back and saddle rich yellow.   Wing bow rich yellow.   Wing bars and primaries black with blue sheen. Secondaries white outer edge, black on tips and inner edge, only white to show when wing is closed.   Remaining plumage black with blue sheen.

Plumage, Female: Hackle silvery white, finely striped with black. Breast salmon, shading off to ash-grey on thighs.   Tail black, but with top feathers matching body plumage.   Remainder French grey or steel grey, very finely pencilled with black and of even shade throughout.

In both sexes: Beak dark horn.   Eyes ruby red.   Comb, face, ear-lobes and wattles red.   Legs and feet willow.

Plumage, Silver Duckwing Variety, Male: Hackle, back, saddle, shoulder coverts and wing bows silver-white.   Secondaries pure white on outer edge, black inner edge.   Only white to show when wing is closed.   Remaining plumage lustrous blue-black.

Plumage, Female: Hackle silver-white, finely striped with black. Breast pale salmon, shading off to ash-grey on thighs. Tail black but with top feathers matching body plumage. Remainder light French grey with very delicate black pencilling.

In both sexes: Beak dark horn. Eyes ruby red. Comb, face, ear-lobes and wattles red. Legs and feet willow.

Self colours are now seldom seen. Whites were standardised with red eyes and yellow legs. Blacks were permitted to have either red or dark purple faces and black legs. Blues had red faces, red eyes and blue-black legs.

Weights: Male 20 to 22 oz. Female 16 to 18 oz. These weights are a little higher than pre-war standards.

SCALE OF POINTS

COLOUR	20
HEAD AND NECK	10
TAIL	10
CONDITION AND SHORTNESS OF FEATHER	10
TYPE AND STYLE	30
LEGS AND FEET	10
EYES	10
	100

Serious Defects: Eyes other than standard for the variety. Flat shins. Crooked breast. Turkey-breast. Wry tail. Roach back. Duck feet. Twisted toes. Any other deformity.

OLD ENGLISH GAME

THIS standard is compiled from that of the O.E.G. Bantam Club, and follows the Oxford ideal. Other standards exist, but essential differences are slight. Chief variations are in methods of interpretation. O.E.G. bantams are of comparatively recent creation. They were evolved largely from the common cross-bred bantam of the countryside. Probably there is very little large breed blood in them.

In the large breed it is usually agreed that a good Game bird cannot be a bad colour. This remark does not apply to bantams, which are show birds only, colour playing a very important part. Nevertheless, the ideal is that the bantam should be a true miniature of the National fighting Game—though this is seldom the case.

GENERAL CHARACTERISTICS: MALE

Carriage: Bold, sprightly, defiant and proud, active on feet, agile, quick in movement, ready for any emergency.

Type: Body short. Back flat, broad at shoulders, tapering to tail. Breast broad, full and prominent, with strong pectoral muscles and breast bone not deep or pointed. Belly small and tight. Wings large, long and powerful, with strong quills amply protecting the thighs. Tail large, carried upwards and spread, quills and main feathers large and strong.

Head: Small and tapering, skin of face and throat flexible and loose. Beak big and boxing (the upper mandible shutting tightly over the lower), crooked or hawk-like, pointed, strong at base, Eyes large, bold, fiery and alike in colour. Comb single, small. upright and thin. Ear-lobes and wattles fine, small and thin. (It is customary to dub cocks. Lobes and wattles are trimmed clean, but the comb is not dubbed so closely as in Modern Game.)

Neck: Large-boned, round, strong and fairly long, neck hackle long and wiry and covering the shoulders.

Legs and Feet: Legs strong, clean-boned, sinewy, close-scaled, not fat and gummy, not stiffly upright, not too wide apart, with a good angle at the hock. Thighs short, round and muscular, following line of body, slightly curved. Toes, four, long, straight and tapering, with long curved nails; hind toe of good length and strength, straight and firm on ground. Spurs hard and fine, set low.

Plumage: Hard, resilient, smooth and glossy, without much fluff.

Handling: Well balanced, mellow, firm and corky, hard yet light-fleshed, with strong contraction of wings and thighs to body.

FEMALE

The general characteristics are similar to those of the male, allowing for the natural sexual differences. The tail is inclined to fan-shape and carried well up.

COLOUR

Colours are very numerous, the most popular being Spangles, Black-reds (Wheaten-bred and Partridge-bred), Duckwings, Brown-reds, Self-blacks and Blues, Furnesses, Creles and Greys. Most of these main colours follow the usual colour-pattern applicable to the variety.

Old English Game
Black Male
Wheaten Female

Furnesses are black with brassy hackles, wing shoulders and backs in males; and black with greyish-brown streaked breasts and wings in females.

Creles are barred varieties of other colours, distinct from Cuckoos, which are plain barred grey.

In Black-reds, the Wheaten-bred is brighter in colour and more popular than Partridge-bred.

Off colours are very numerous, but not so popular as in the large breed. Piles and Whites are seldom seen.

Hennies (hen-feathered males) muffs and tassels (beards and top-knots) are recognised sub-varieties in all colours.

Colour of face, beak, eyes, shanks and toe-nails varies with the particular colour-variety; from red face, red eyes, white legs, toe-nails and beak in Spangles to deep crimson, purple or black faces, black eyes, black legs and feet in Brown-reds and similar dark colours. Willow, olive and yellow legs are frequent and permissible in various sub-varieties.

Some of the more popular colour-varieties, with the names usually applied to them by showmen, are given below.

Plumage, Black-red Variety (Partridge-bred), Male: Cap dark red, neck hackle and saddle hackle dark rich red shading to deep orange. Back and saddle deep crimson. Breast, thighs, belly and tail black. Wing bars steel blue, secondaries showing bay when closed.

Plumage, Female: Hackle golden colour lightly striped with black. Breast salmon to robin red, shading off to ash-grey on belly and thighs. Back, shoulders and wings even partridge brown stippled with fine darker markings. Primaries and tail dark partridge to black.

In both sexes: Eyes dark red, face blood red. Legs and beak willow, yellow, white or olive. Dark legged birds should have grey fluff, white or yellow legged birds should have light fluff.

Plumage, Black-red Variety (Wheaten-bred), Male: Cap bright red, neck hackle bright orange shading off to bright lemon. Saddle hackle similar. Back and saddle bright crimson. Breast, thighs, belly and tail black. Otherwise similar to Partridge-bred male but much brighter in colour.

Plumage, Female: Wheaten colour (a pale creamy tint resembling wheat) with clear red hackle. Tail and primaries nearly black, the breast and underparts a very delicate creamy self colour. Colour varies from very pale to a dark or red wheaten (the colour of red wheat) with dark red hackles and black tail feather.

Old English Game
Duckwing Male,
Spangled Female

In both sexes: Eyes fiery red, face bright red. Legs and beak white, occasionally yellow. Fluff white.

Plumage, Yellow (Golden) Duckwing Variety, Male: Face red. Hackle yellow, saddle straw colour, shoulders golden. Wing bars steel blue. Secondaries white when closed. Remaining plumage black.

Plumage, Female: Hackle white lightly striped with black. Body and wings deep silvery grey. Breast deep salmon shading off to ash-grey on thighs. Primaries and tail nearly black.

In both sexes: Eyes pearl or red, beak and legs yellow, white or dark. Fluff light grey.

Plumage, Silver Duckwing Variety, Male: Breast, thighs, belly and tail black. Wing bars steel blue. Hackles, saddle, shoulders clear silvery white. Secondaries white when closed.

Plumage, Female: White hackle lightly striped with black. Body and wings even silvery grey. Breast pale salmon. Primaries and tail nearly black.

In both sexes: Eyes pearl or red. Beak and legs white or dark. Eye colour light or dark according to leg colour.

Plumage, Brown-red Variety, Male: Hackles, back and wing bow orange to lemon. Remainder green-black, the breast feathers edged with orange as far down as top of thighs. Colour generally deeper than in Modern Game.

Plumage, Female: Cap orange, neck hackle light orange. Remainder green-black, breast feathers delicately laced with orange as in male. Back free from lacing and shoulders free from ticking.

In both sexes: Eyes black, beak and legs black, faces deep purple-black.

Plumage, Spangled Variety, Male: Closely resembles the Partridge-bred Black-red variety but with ends of feathers regularly tipped with small white spangles. The more regularly spangles are distributed the better. Neck and saddle hackle should be similarly tipped.

Plumage, Female: Closely resembling the Partridge-bred Black-red female but each feather finished with a small white spangle tip—the more evenly distributed the better.

In both sexes: Eyes dark red, face blood red. Legs and beak usually white, occasionally yellow.

Note: In O.E.G. bantams spangles are always of Partridge-bred colouring. Spangles are not known in Wheaten-bred Black-reds.

Self Colours: *Blacks* and *Blues* are very popular. They should be free from white or coloured feathers, but may have red eyes, red faces and white legs, or black eyes, purple-black faces and black legs and beaks. *Whites* are seldom seen in bantams. They should be free from coloured feathers, with eyes red, faces red, legs and beaks white or yellow.

Although colour is less important in Old English Game than in Moderns, general colour descriptions in the main varieties are similar.

Suggested Weights: Male 22 to 26 oz. Female 18 to 22 oz. These weights show considerable increases on pre-war standards.

SCALE OF POINTS

HEAD, BEAK AND EYES	10
BODY, BREAST, BACK AND BELLY	20
TAIL	6
SHANKS, SPURS AND FEET	10
HANDLING, HARDNESS, CONDITION AND CONSTITUTION	15
NECK	6
WINGS	7
THIGHS	8
PLUMAGE AND COLOUR	9
CARRIAGE AND ACTION	9
	100

Serious Defects: Thin thighs or neck. Deep keel. Pointed, crooked or indented breast bone. Flat sides. Stork legs. Duck-feet. In-knees. Soft flesh. Thick insteps or toes. Bad carriage. Bad action. Broken, soft or rotted plumage. Indications of weak constitution.

ORPINGTON

In breed character and feather Orpington Bantams are among the most typical of all miniatures of large breeds, even though there is a tendency for some strains to be somewhat oversized. Whites are probably the most popular.

GENERAL CHARACTERISTICS : MALE

Carriage: Bold, erect, well balanced and active.

Type: Body deep, broad and cobby. Breast broad, full and well rounded. Saddle wide, slightly rising to tail, giving the back a somewhat short and concave outline. Wings compactly carried. Tail compact, broad, somewhat short, carried somewhat high, but by no means " squirrel ", with well curved sickles.

Head: Small. Beak short, strong and slightly curved. Eyes full, round, and bright. Comb single, small, straight, firmly set, and evenly serrated. Face smooth and fine. Ear-lobes small and elongated. Wattles small, elongated and well rounded at the tips.

Neck: Medium length, well curved, abundantly feathered and well covered with hackles.

Legs and Feet: Legs short, strong and set well apart. Thighs almost hidden by body feathering. Shanks straight, stout but not coarse, close scaled, free from feathers. Toes, four straight and well spread.

Plumage: Profuse, broad and soft, but fairly close not loose; avoiding hardness.

Handling: Firm.

Skin: Fine, silky.

FEMALE

The general characteristics are similar to those of the male, allowing for the natural sexual differences.

COLOUR

Plumage, Black Variety, Male and Female: Black with brilliant beetle-green sheen; undercolour dense black.

In both sexes: Beak black, eyes black with dark brown iris. Shanks and toes black. Toe-nails, soles of feet and skin white. Comb, face, wattles and ear-lobes bright red.

Plumage, Buff Variety, Male and Female: Clear, sound, even buff throughout to the skin.

In both sexes: Beak white or horn, eyes orange or red. Shanks, feet and toe-nails white, with or without pinkish tinge. Skin white. Comb, face, wattles and ear-lobes bright red.

Plumage, White Variety, Male and Female: Pure, snow white.

In both sexes: Beak and skin white. Eyes red. Shanks, feet and toe-nails white, with or without pinkish tinge. Comb, face, wattles and ear-lobes bright red.

Plumage, Blue Variety, Male: Hackles, saddle, wing bow, back and tail dark slate blue. Remainder medium slate blue, each feather to show lacing of darker shade as on back.

Plumage, Female: Medium slate blue, laced with darker shade all through except head and neck, which are a dark slate blue.

In both sexes: Beak blue, eyes hazel or dark brown. Legs and feet black or blue. Comb, face, wattles and ear-lobes bright red.

Standard Weights: Male 32 to 36 oz. Female 28 to 32 oz.

SCALE OF POINTS

	Black	Blue	Buff and White
TYPE, CARRIAGE AND FEATHER	35	30	30
COLOUR AND UNDERCOLOUR	15	30	20
HEAD AND EYES	20	15	15
LEGS, FEET AND SKIN	15	10	10
SIZE AND CONDITION	15	15	25
	100	100	100

Serious Defects: Yellow skin. Yellow in shanks or feet. White in ear-lobe, feathered shanks or feet. Willow shanks. Light eyes. Side sprigs on comb. Squirrel or wry tail. High on legs, flat back. Square breast. Split wing. Any deformity. Lack of condition.

PLYMOUTH ROCK

THESE have little or no large breed blood. The Barred variety was probably evolved largely from Scots Greys in the first place. There have been great advances in type and colour. Three colours (Barred, Buff and Partridge) are standardised. Blacks and Whites are sometimes seen but are usually either sports from the Barred variety or single-combed Wyandottes.

GENERAL CHARACTERISTICS: MALE

Carriage: Upright and smart, well balanced, free from stiltiness.

Type: Body large, deep and compact, evenly balanced and symmetrical, broad, keel bone long and straight. Back broad and of medium length; saddle feathers medium length and abundant. Breast full, deep, broad and well rounded. Wings medium size, well tucked up, ends covered by saddle hackles. Tail medium size, slightly rising, inclined backwards; sickles curved and medium length, with abundant coverts hiding the main tail feathers.

Head: Medium size, strong and carried well up. Beak short, stout and slightly curved. Eyes bright and clear. Comb single, straight and erect, with well-defined serrations free from side sprigs. Ear-lobes of fine texture, showing no white. Wattles moderately rounded and of equal length.

Neck: Slightly curved, medium length, with profuse hackle flowing well over shoulders.

Legs and Feet: Legs well apart. Thighs large and medium length. Shanks medium length, straight and stout, free from feathers. There must be no tendency to knock-knees. Toes, four, strong, straight and well spread.

FEMALE

The general characteristics are similar to those of the male, allowing for the natural sexual differences.

COLOUR

Plumage, Barred Variety, Male and Female: White ground with blue tinge, each feather barred across with black bands having a beetle-green sheen, the bands moderately narrow and equal in breadth, colours sharply defined and not blurred or shaded off.

266

Plymouth Rock
Buff Male

Plymouth
Rock
Barred
Female

Barring should continue through the shaft and into under-fluff, and each feather must finish with a black tip. Plumage should present a bluish appearance, free from brassiness and a uniform shade all over.

Plumage, Buff Variety, Male and Female: Clear, sound even buff, of one golden shade throughout. Tail, wings and fluff to harmonise. Feathers should not show lacing, peppering or mealiness.

In both sexes: Beak, legs and feet bright yellow. Eyes rich red or bay. Comb, face, wattles and ear-lobes bright red.

Plumage, Partridge Variety, Male: Head bright red. Neck hackles, web of feather, solid, lustrous, greenish black, of moderate width, with a narrow edging of a medium shade of rich brilliant red, uniform in width, extending around point of feather; shaft black; plumage on front of neck black, as clear as possible of red. Wings: fronts, black; bows, a medium shade of rich brilliant red; coverts, lustrous, greenish black, forming a well-defined bar of this colour across wing when folded; primaries, black, lower edges, reddish bay; secondaries, black, outside webs, reddish bay, terminating with greenish-black at the end of each feather, the secondaries when folded forming a reddish bay wing-bay between the wing-bar and tips of secondary feathers. Back and saddle a medium shade of rich, brilliant red, with lustrous, greenish-black stripe down the middle of each feather, same as in hackle. A slight shafting of rich red is permissible. Tail black; sickles and smaller sickles, lustrous, greenish black; coverts, lustrous, greenish-black, edged with a medium shade of rich, brilliant red. Body black; lower feathers slightly tinged with red; fluff black, slightly tinged with red. Breast lustrous black (slight tinge of red allowed). Lower thighs black. Undercolour of all sections, slate.

Plumage, Female: Head deep reddish bay. Neck reddish bay, centre portion of feathers black, slightly pencilled, with deep reddish bay; feathers on front of neck same as breast. Wings: shoulders, bows and coverts, deep reddish bay with distinct pencillings of black, outlines of which conform to shape of feathers; primaries, black with edging of deep reddish bay on outer webs; secondaries, inner web, black, outer web, deep reddish bay with distinct pencillings of black extending around outer edge of feathers. Back deep reddish bay with distinct pencillings of black, the outlines of which conform to shape of feathers. Tail black, the two top feathers pencilled with deep reddish bay on upper edge; coverts deep reddish bay pencilled with black. Body deep reddish bay pencilled with black; fluff deep reddish bay. Breast deep reddish bay with distinct pencillings of black, the outlines of which conform to shape of feathers. Lower thighs deep reddish bay pencilled with black. Undercolour slate.

Note: Each feather in back, breast, body, wing bows and thighs to have three or more distinct pencillings. In both sexes: Beak yellow. Eyes reddish bay. Legs and feet yellow. Comb, face, wattles and ear-lobes bright red.

Suggested Weights: British Bantam Association—Male 26 to 30 oz. Female 22 to 26 oz. The standards of the Buff Rock Club and the Barred Rock Bantam Club are respectively slightly below and above these figures.

SCALE OF POINTS

Barred Variety

TYPE	20
COLOUR	20
BARRING	20
LEGS AND FEET	10
HEAD	5
TAIL	5
SIZE	10
CONDITION	10
	100

Buff Variety

TYPE	20
CARRIAGE AND SIZE	10
COLOUR	20
QUALITY AND TEXTURE	15
HEAD	10
EYE COLOUR	5
LEGS AND FEET	5
CONDITION	15
	100

Partridge Variety

TYPE	30
COLOUR AND PENCILLING	30
HEAD AND EYES	10
LEGS AND FEET	10

Continued on next page

Plymouth Rock Scale of Points—continued

       CONDITION 10
       QUALITY AND TEXTURE.................. 10

                                                       100

Serious Defects: Indications of feather or fluff on shanks or feet. Shanks other than yellow. White ear-lobes. Pale or faulty-coloured eyes. Slipped wing. Foul-coloured feathers in the Barred variety. Spotted, laced or mealy plumage in Buffs. Black or white in wings or white in tail in Buffs. Any deformity.

POLAND OR POLISH

THESE are miniatures of the old-established big breed. Whites are smallest in size and best in quality at present, some of the other varieties being oversized.

GENERAL CHARACTERISTICS: MALE

Carriage: Erect and sprightly.

Type: Body full, round and fairly long. Back flat, tapering to tail; shoulders wide, flanks deep. Wings large, closely carried. Tail full and neatly spread but carried rather low. Sickles and coverts abundant and well curved.

Head: Large, with a decided and pronounced protuberance on top, from which springs a large full crest, circular on top, free of split or parting, compact in centre and falling evenly all round with untwisted feathers similar to the hackle feather. Beak medium, with large and very prominent nostrils rising above the curved line of beak. Eyes large and full. Comb (if any) horn type and very small; preference normally being given to birds without combs. Face smooth, beardless in white-creasted colour-varieties, completely covered by muffling in other colours. Muffling plentiful, full and compact, fitting round and almost hiding the face. Ear-lobes very small and round, unseen in muffed varieties. Wattles of fair size in white-crested colour-varieties. No wattles in other colours.

Neck: Long, with abundant hackle covering the shoulders.

270

Poland, White Male,
White-Crested
Black Female

Legs and Feet: Legs fairly long and slender.   Shanks free of feathers. Toes, four, slender and well spread.

FEMALE

The general characteristics are similar to those of the male, allowing for the natural sexual differences.   The crest, however, must be globular.

COLOUR

Plumage, Black: Rich metallic black throughout.

In both sexes: Beak, eyes, face, legs, etc., as in Whites.

Plumage, Chamois or White Laced Buff Variety, Male: An even shade of buff ground colour throughout, with white lacing round edges.   Crest to be white at root and tips, but as free as possible from all-white feathers.   Muffling to be mottled or laced, not solid buff colour.   Hackles tipped with white, wing bar and secondaries laced with white, primaries tipped.   Tail, sickles and tail coverts laced with white.

Plumage, Female: Except that wing primaries are tipped with white, the whole remaining plumage, including crest, should be buff ground colour with white lacing.

In both sexes: Beak dark blue or horn.   Eyes, comb and face red.   Ear-lobes blue-white.   Legs and feet dark blue.

Plumage, Gold Variety, Male: Golden bay ground colour with black markings.   Crest black at root and tips and as free as possible of all-white feathers.   Muffling mottled or laced, not solid black or gold.   Hackle tipped with black.   Back and saddle laced or spangle-tipped with black.   Breast, thighs, shoulders and wings laced black.   Wing primaries tipped with black.   Tail laced and ends of sickles splashed black.

Plumage, Female: Golden bay ground throughout with black lacing, each feather distinctly marked and as free as possible from splashes.

In both sexes: Beak, eyes, lobes, legs, etc., as in the Chamois.

Plumage, Silver Variety, Male and Female: As in the gold, substituting silver for gold as ground colour throughout.

In both sexes: Beak, legs, etc., as in the Chamois.

White in crest of laced varieties is usual and permissible in birds over one year old, and should not be penalised.

It is popularly (but wrongly) supposed that white-crested colour-varieties should have clear white crests, as described in Poultry Club Standards up to the 8th edition.   This is not the case.   In these varieties there should be broad bands of coloured feathers at base of crest and above the beak, composed of feathers the same colour as body-feathers. Birds without these bands should be passed. It means they have been plucked.

Plumage, White Variety, Male and Female: Pure white throughout.

In both sexes: Beak dark blue. Ear-lobes white. Eyes, comb and face red. Legs and feet dark blue.

Plumage, White-crested Black Variety, Male and Female: Rich metallic black, except the crest, which is snow white with a black band at base in front of face.

In both sexes ; Beak, eyes, face, legs, etc., as in Whites.

Plumage, White-crested Blue Variety, Male and Female: Solid dark blue (self coloured). Crest snow white with a blue band at base in front of face.

In both sexes: Beak, eyes, face, legs, etc., as in Whites.

Suggested Weights: Male 24 to 28 oz. Female 18 to 24 oz. The weights of some colours are often considerably more.

SCALE OF POINTS

White-crested colour-varieties

CREST	30
HEAD	15
COLOUR	25
TYPE	5
SIZE	15
CONDITION	10
	100

Other colours

CREST	30
HEAD AND MUFFLING	15
COLOUR AND MARKINGS	25
TYPE	5
SIZE	15
CONDITION	10
	100

Serious Defects: Split or twisted crest. Comb (if any) other than horn type. Absence of muffling in Chamois, Gold, Silver and White varieties. Legs other than blue or slate. Any deformity.

RHODE ISLAND RED

T HERE remain points of difference between the Rhode Island Red Club and the Rhode Island Red Bantam Club. The standard which follows is a compromise between the two.

GENERAL CHARACTERISTICS: MALE

Carriage: Active, alert and well-balanced.

Type: Body broad and long, fair depth, distinctly oblong rather than square, with long straight keel bone extending well forward and back. Back broad, long and horizontal. Breast broad and full, carried nearly perpendicularly in line with base of beak. Wings large and well folded, with horizontal flights. Tail of moderate size (this is important), carried well back, the sickles a little longer than the main tail feathers, well spread and carried at a low angle (but not drooping) thus increasing the apparent length of the body. Feathering close, fluff moderately full.

Head: Medium size, carried slightly forward. Beak curved, moderate length. Eyes large, prominent and bright. Comb single, medium size, fine texture, erect, straight and firmly set, with five even serrations. (A rose-combed variety is standardised in the large breed, but seldom seen in bantams.) Face smooth. Ear-lobes fine and well developed. Wattles medium size and moderately rounded, of fine texture.

Neck: Medium length, profusely hackled but not loose-feathered; carried slightly forward.

Legs and Feet: Legs of medium length. Thighs large and well feathered; shanks well rounded and free of feathers. Toes, four, strong, straight and well spread.

FEMALE

The general characteristics are similar to those of the male, allowing for the natural sexual differences.

COLOUR

Plumage, Male: Hackle red, without black markings, matching body colour. Wing primaries, lower web black, upper red; secondaries, lower web red, upper black; flight coverts black;

274

wing bows and coverts red. Main tail and sickles black or green-black. Coverts mainly black, but red approaching the saddle. Remainder rich brilliant red, free from shaftiness, mealiness, peppering or " ginger ". The bird should be of very brilliant lustre and have a glossed appearance. Undercolour and quill colour red or salmon, without smut or white. Black or white in undercolour is most undesirable; and (other things being equal) the richest red undercolour shall carry the award. (*Note :* The " red " colour nowadays favoured is an extremely deep chocolate-red, and though some breeders disagree with this description, few birds of lighter colour receive prizes.)

R.I.R. Club judges are now instructed to pass " over-prepared " birds without comment, and to adhere strictly to the colour-standard, which calls for rich brilliant red.

Plumage, Female: Hackle red, tips of lower feathers having black ticking but not heavy markings. Tail black or green-black. Remainder generally as described for the male, except that the female will not be so lustrous.

In both sexes: Beak, legs and feet red-horn or yellow. Eyes red; comb, face, ear-lobes and wattles brilliant red.

Weights: Male 28 to 32 oz. Female 24 to 28 oz. The Bantam Club favours lower weights.

SCALE OF POINTS

COLOUR OF PLUMAGE	25
SIZE	25
TYPE	15
CONDITION	10
EYE COLOUR	9
LEGS	6
HEAD AND COMB	10
	100

(*Note :* This scale of points is the one adopted by the Bantam Club. The scale followed by the Rhode Island Red Club approximates closely to that for larger Rhodes.)

Serious Defects: Feather or down on shanks, or indications of plucking same. Lopped combs or side sprigs. Wall eyes. White showing in outer plumage. Lobes more than half-white. Shanks and feet other than yellow or red-horn. Any deformity.

(*Note:* To these serious defects should be added Leghorn type tails and the frizzled, Silkie type or otherwise defective feather often developed through concentration on lustrous dark plumage.)

ROSECOMB

THE Rosecomb bantam is a gem of show birds. In former days it achieved probably the highest pitch of artificial perfection ever achieved in exhibition birds.

GENERAL CHARACTERISTICS: MALE

Carriage: Cobby but not dumpy. The back should show one sweeping curve from neck to sickles.

Type: Body short and broad. Back short, shoulders broad and flat. Breast carried well up and forward, with a bold curve from wing bow to wing bow. Wings carried rather low, showing only front half of thighs. Wide flight feathers round-ended and broad to ends. Stern flat, broad and thick (not running off to nothing at setting-on of tail), with abundant feather; the saddle hackle long and plentiful and extending from tail to middle of back. Tail carried well back, main feathers broad and overlapping neatly; the sickles being long, circled with a bold sweep, broad from base to rounded ends, main tail feathers not projecting beyond the sickles. Furnishing feathers plentiful, broad from base to end, round-ended and uniformly curved with the sickles but hanging somewhat shorter; side hangers broad and long and with the hackles filling the space between stern and wing ends. All feather broad to ends.

Head: Short and broad. Beak stout and short. Comb rose, neat and long, with square well-filled front, set firmly, tapering to the setting-on of the spike or leader; top perfectly level and crowded with small round spikes. The leader stout at base, firm, long and perfectly straight, tapering to a fine point. Comb and leader rise slightly from front to rear in one line. Face of fine texture. Ear-lobes absolutely round, with rounded edges, of uniform thickness all over, not hollow or dished, firmly set on the face and kid-like in texture; not smaller than a sixpence ($\frac{3}{4}$ in) or larger than a shilling ($\frac{7}{8}$ in). Wattles round, neat and fine.

Neck: Rather short, well curved, with wide feathers, the hackle

Rosecomb
Black Male
& Female

falling gracefully and plentifully over shoulders and wing bows and almost reaching the tail.

Legs and Feet: Legs short. Thighs set well apart, stout at top and tapering to hocks. Shanks rather short, round, fine and free of feathers. Toes, four, straight and well spread.

FEMALE

With the exception of the ear-lobes, which should not be larger than the now defunct silver threepenny piece (approx. $\frac{5}{8}$ in), and the wings, which are not carried so low but are not tucked up, the general characteristics are similar to those of the male, allowing for the natural sexual differences.

(*Note:* Standard sizes of ear-lobes are usually considerably exceeded in show specimens.)

COLOUR

Plumage, Black Variety, Male and Female: Black with brilliant green sheen from head to end of tail, the wing bar with extra bright green sheen. Tail feathers and sickles to be rich in green sheen. Beak black, eyes hazel or brown. Legs and feet black.

Plumage, Blue Variety, Female: Blue of medium shade, free from lacing. The plumage of hackles, back and shoulders in males of a darker shade.

Plumage, White Variety, Male and Female: Snow-white, free from straw tinge. Beak white, eyes red. Legs and feet white.

In both sexes and varieties: Comb, face and wattles brilliant cherry red. Ear-lobes spotlessly white, especially near wattles.

Weights: Male 20 to 22 oz. Female 16 to 18 oz.

SCALE OF POINTS

HEAD (COMB 20, LOBES 15)	35
TAIL	15
COLOUR	12
TYPE	15
CONDITION	15
LEGS, ETC.	8
	100

Serious Defects: Stiltiness. Narrow chest or back. Hollow-fronted or leafy comb. Coarse bone. Tightly carried wings. Narrow feathers. Blushed lobes. Coloured feathers. White in face. In Blacks, grizzled or brown flights; purple sheen or barring; light legs.

SCOTS GREY

THESE were reasonably popular up till about 1930, and were much the vogue in Scotland.   They were the smallest of the barred varieties of bantams, and might well be revived in popularity.   They are now seldom seen.

GENERAL CHARACTERISTICS: MALE

Carriage: Erect, active and bold.

Type: Body compact, fairly long.   Back flat and broad.   Breast full and carried high.   Wings of moderate length, well tucked up, bow and tips covered by hackles.   Tail fairly long and well sickled, carried well up.

Head: Long and fine.   Beak curved and strong.   Comb single, upright, medium size, well serrated, blade following line of skull. Eyes large and bright.   Face fine textured.   Ear-lobes and wattles medium size.

Neck: Tapered and with profuse flowing hackle.

Legs and Feet: Legs longish and strong.   Thighs well apart but not as prominent as in Game fowl.   Shanks clear of feathers. Toes, four, stout, strong, straight and well spread.

Handling: Firm, similar to Game.

FEMALE

With the exception that the comb may be either erect or slightly drooping, the general characteristics are similar to those of the male, allowing for the natural sexual differences.

COLOUR

Plumage, Male: Cuckoo-feathered.   Ground colour blue-white, but may vary to light grey on neck, saddle and tail so long as the shade and depth of colour register the same throughout.   Barring black with a metallic lustre, straight across normal body-feather but slightly V-shaped on hackles and tail.   Barring and ground colour to show bands equal in width.   Plumage free from red, black, white or yellow feathers, and all markings small and sharply defined, with the tail distinctly and evenly barred.

Plumage, Female: Similar, except that the markings are not so small, and the general appearance of barring is less regular.

In both sexes: Beak white, or white streaked with black. Comb, face, wattles and ear-lobes bright red. Eyes amber. Legs and feet white, or white mottled with black (not sooty or dusky).

Weights: Male 22 to 24 oz.   Female 18 to 20 oz.

SCALE OF POINTS

COLOUR AND MARKINGS (HACKLE 10, BACK 10, TAIL 10, WINGS AND SHOULDERS 10, BREAST AND THIGHS 10)   50
SIZE...   15
TYPE...   10
HEAD...   10
CONDITION..   10
LEGS AND FEET....................................   5
                                                  ———
                                                   100

Serious Defects: Distinct characteristics of any other breed.   Any deformity.

SEBRIGHT

THIS breed is a genuine bantam and one of the oldest British varieties. It has no counterpart in large breeds, but has played a part in the production of other laced fowl, notably Wyandottes. There are two colours, Gold and Silver.

GENERAL CHARACTERISTICS: MALE

Carriage: Strutting and tremulous, on tip-toe, somewhat resembling a Fantail pigeon.

Type: Body compact, with broad and prominent breast. Back very short. Wings large and carried low. Tail square, well spread and carried high. Sebright males are hen-feathered, without curved sickles or pointed neck and saddle hackles.

Head: Small. Beak short and slightly curved. Comb rose, square-fronted, firmly and evenly set on, top covered with fine

D.P.S. 11

points, free from hollows, narrowing behind to a distinct spike or leader, turned slightly upwards. Eyes full. Face smooth. Ear-lobes flat, and unfolded. Wattles well rounded.

Neck: Tapering, arched and carried well back.

Legs and Feet: Legs short and well apart. Shanks slender and free from feathers. Toes, four, straight and well spread.

Plumage: Short and tight, feathers not too wide but never pointed. (Almond-shaped feather is desired.)

FEMALE

The general characteristics are similar to those of the male, allowing for the natural sexual differences. Her neck is upright.

COLOUR

Plumage, Gold Variety, Male and Female: Uniform golden bay with glossy green-black lacing and dark grey undercolour; each feather evenly and sharply laced all round its edge with a narrow margin of black. Shaftiness is undesirable.

Plumage, Silver Variety, Male and Female: Similarly marked on pure, clear silver-white ground colour.

In both sexes: Beak dark horn in Golds; dark blue or horn in Silvers. Eyes black, or as dark as possible. Comb, face, wattles and ear-lobes dark purple or dull red (mulberry). Legs and feet slate-blue.

Although in males the purple or mulberry face is seldom obtainable, the eye should be dark and surrounded with a dark cere.

Weights: Male 22 oz.   Female 18 oz.

SCALE OF POINTS

LACING	25
COMB	5
FACE AND LOBES	10
GROUND COLOUR	15
TAIL	10
TYPE	20
WEIGHT	5
CONDITION	10
	100

There is at present a decided move on foot to improve type and to discourage the prevailing whip tails and narrow build, particularly in females.

Serious Defects: Single comb. Sickle feathers or pointed hackles on the male. Feathers on shanks. Legs other than slate-blue. Other than four toes. Any deformity.

SUSSEX

THESE are seen in all the colours of the Large Sussex. No separate bantam club exists, but bantams are sponsored by the large breed club, which wisely lays great stress on type.

GENERAL CHARACTERISTICS: MALE

Carriage: Graceful, showing length of back, vigorous and well balanced.

Type: Shoulders wide. Back broad and flat. Breast broad and square, carried well forward, with long straight breast bone. Wings carried close. Tail of moderate size carried at an angle of about 45 degrees.

Head: Medium size. Beak short, strong and curved. Comb single, medium, upright, evenly serrated, fitting close to head. Eyes full, prominent and bright. Face smooth and of good texture. Ear-lobes and wattles medium size and fine texture.

Neck: Medium length, with fairly full hackles.

Legs and Feet: Thighs short and stout. Shanks short and strong, rather wide apart, free from feathers, closely scaled. Toes, four, straight and well spread.

Plumage: Close and free from unnecessary fluff.

FEMALE

The general characteristics are similar to those of the male, allowing for the natural sexual differences.

COLOUR

Plumage, Brown Variety, Male: Head and neck hackles rich dark mahogany striped with black. Saddle hackle same as neck hackle. Back and wing bow rich dark mahogany. Wing coverts forming the bar, blue-black; secondaries and flights black, edged with brown. Breast, tail and thighs black.

Plumage, Female: Head and neck hackles brown striped with black. Back and wings dark brown, finely peppered with black. Breast and underbody clear pale wheaten brown. Flights black, edged with brown. Tail black.

Plumage, Buff Variety, Male and Female: Similarly marked to the Light, but with rich even golden-buff ground colour. Buff under-colour is desirable, but dark colour is not at present penalised.

Plumage, Light Variety, Male and Female: Head and neck hackle white striped with black, the solid black centre of each feather entirely surrounded by a white margin. Wings white with black flight markings. Tail coverts black, tail black. Remainder of plumage pure white throughout.

Plumage, Red Variety, Male and Female: One uniform shade of rich dark red, with black striped neck hackle, black in flights, black tail and slate undercolour.

Plumage, Silver Variety, Male: Head, neck and saddle hackles silver-white striped with black, the black striping entirely surrounded by a white margin. Back and wing bow silvery-white. Wing bar black. Flights and secondaries black with grey markings. Breast black, with silver lacing round feathers and white shafts. Thighs dark grey faintly laced with silver. Tail black. Undercolour grey, shading to white at skin.

Plumage, Female: Head and neck hackles as described for male. Back and wing bow grey-black, each feather with white shaft and fine silver edging. Flights and secondaries greyish-black. Tail black. Breast and thighs a lighter shade of grey-black with white shafts and silver lacing. Undercolour as described for male.

Plumage, Speckled Variety, Male and Female: Head and neck hackle rich dark mahogany striped with black and tipped with white. Saddle hackle similar. Wing primaries white, brown and black, white not predominating. Main tail feathers black splashed with white, sickles black with white tips. Remainder of plumage rich dark mahogany, each feather tipped with a small white spot, a narrow glossy black bar dividing the white from the ground colour. Undercolour slate and red with a minimum of white. Mahogany ground colour to be free from peppering.

Plumage, White Variety, Male and Female: Pure snow-white throughout.

In both sexes: Beak white or horn. Eyes red in Speckled, Buff or Red varieties; orange in Lights, Silvers and Whites. Comb,

Sussex
Silver Male
and Female

face, wattles and ear-lobes red.   Legs and feet white; flesh and skin white.

Weights: Male 40 oz (max.).   Female 28 oz (max.).

SCALE OF POINTS

TYPE, SIZE AND WEIGHT	35
HACKLE, TAIL AND WING	20
BODY COLOUR	15
HEAD AND EYE	15
FEET AND LEGS	15
	100

Serious Defects: Other than four toes.   Feather on shanks or toes. Comb other than single.   Wry tail or any other deformity.

WELSUMMER

WELSUMMER bantams are typical of their larger prototype and some produce eggs which have the same deep shell colour.   Both the larger fowl and the bantam are sponsored by the one breed club.

GENERAL CHARACTERISTICS: MALE

Carriage: Upright, alert and active.

Type: Body well built on good constitutional lines.   Back broad and long.   Breast full, well rounded and broad.   Wings moderately long, carried closely to the sides.   Tail fairly large and full, carried high, but not squirrel.   Abdomen long, deep and wide.

Head: Symmetrical, well balanced, of fine quality without coarseness, excesses or exaggeration.   Skull refined, especially at back. Beak strong, short and deep.   Eyes keen in expression, bold, full, highly placed in skull and standing out prominently when viewed from front or back; pupils large and free from defective shape. Comb single, of medium size, firm, upright, free from any twists or excess around nostrils, clear of nostrils, and of fine, silky texture, five to seven broad and even serrations, the back following closely but not touching the line of the skull and neck.   Face smooth, open and of silky texture, free from wrinkles or surfeit of flesh and

286

without overhanging eyebrows. Ear-lobes small and almond shaped. Wattles of medium size, fine and silky texture and close together.

Neck: Fairly long, slender at top but finishing with abundant hackle.

Legs and Feet: Thighs to show clear of body without loss of breast. Shanks of medium length, medium bone and well set apart, free from feathers and with soft, pliable sinews, free from coarseness. Toes, four, long, straight and well spread out, back toe to follow in straight line, free from feathers between toes.

Plumage: Tight, silky and waxy, free from excess or coarseness, silky at abdomen and free from bagginess at thighs.

Handling: Compact, firm and neat bone throughout.

FEMALE

The general characteristics are similar to those of the male, allowing for the natural sexual differences. Handling: Pelvic bones fine and pliable; abdomen pliable; flesh and skin of fine texture and free from coarseness; plumage sleek; abdomen capacious, but well supported by long breast-bone and not drooping; general handling of a fit, keen and active layer.

COLOUR

Plumage, Male: Head and neck rich golden brown. Hackles rich golden brown as uniform as possible, free from black striping, yet underparts (out of sight) may show a little striping at present. Back, shoulder coverts and wing bow bright red-brown. Wing coverts black with green sheen forming a broad bar across; primaries (out of sight when wing is closed), inner web black, outer web brown; secondaries, outer web brown, inner web black with brown peppering. Tail (main) black with a beetle-green sheen; coverts, upper black, lower black edged with brown. Breast black with red mottling. Abdominal and thigh fluff black and red mottled.

Plumage Female: Head golden brown. Hackle golden brown or copper, the lower feathers with black striping and golden shaft. Breast rich chestnut red going well down to the lower parts. Back and wing bow reddish-brown, each feather stippled or peppered with black specks (i.e., partridge marking), shaft of feather showing lighter and very distinct. Wing, bar chestnut brown; primaries, inner web black, outer brown; secondaries, outer web brown, coarsely stippled with black; inner web black, slightly peppered

with brown. Abdomen and thighs brown with grey shading.
Tail black, outer feathers pencilled with brown.

In both sexes: Beak yellow or horn. Eyes red. Comb, face,
ear-lobes and wattles bright red. Legs and feet yellow. Undercolour dark slate grey.

Standard Weights: Male 36 oz.   Female 28 oz.

SCALE OF POINTS

GENERAL TYPE	20
HANDLING, SIZE AND INDICATIONS OF PRODUCTIVENESS	30
HEAD	10
LEGS AND FEET	10
COLOUR	20
CONDITION	10
	100

Serious Defects: Comb other than single or with side sprigs. White
in lobe. Feather on legs, hocks or between toes. Other than four
toes. Striping in neck hackle or saddle of male. Absolutely
black or whole red breast in the male.   Salmon breast in the female.
Legs other than yellow. Badly crooked or duck toes. Any body
deformity. Coarseness, beefiness and anything which interferes
with the productiveness and general utility of the breed.

WYANDOTTE

EXCEPT for O.E. Game and Belgian Bearded Bantams, probably
no other breed has been produced in so many colours as the
Wyandotte. Except in laced varieties there is, of course, very little
big breed blood in their make-up, some of which present extraordinary colour schemes.

GENERAL CHARACTERISTICS: MALE

Carriage: Well balanced and graceful, active and alert but docile
resembling the Brahma.

type: Body short and deep, well rounded at sides.   Back short, with

288

Wyandotte
Columbian Male
Laced Female

broad full saddle rising to tail with concave sweep. Breast full round, with a straight keel. (To the eye the body should give an impression of greater depth than length.) Wings medium size, closely folded to the side. Tail full but short, well developed and spread at base; the true tail feathers carried rather upright, sickles medium length. (The main tail feathers should be rather softer than in most breeds but not Cochiny.) General impression of the body to suggest curves in all directions.

Head: Broad and short. Beak stout and well curved. Comb helmet or rose, firm on head, square and low in front, tapering evenly towards the back and ending in a well-defined leader following curve of neck without any upward tendency. Top of comb covered with small rounded points, all outlines convex and conforming to shape of skull. Face smooth and fine. Ear-lobes oblong and well developed. Wattles medium length, fine and rounded.

Neck: Well arched, medium length, with full hackle.

Legs and Feet: Legs of medium length. Thighs well covered with soft feathers, the fluff abundant but fairly close and silky. Shanks strong, well rounded and free of feather or fluff. Toes, four, straight and well spread.

Plumage: Fairly close and silky, with fluff abundant but not long. (Wyandotte feathering should give an impression of abundance rather than length.)

FEMALE

The general characteristics are similar to those of the male, allowing for the natural sexual differences.

COLOUR

Plumage, Black Variety, Male and Female: Sound black throughout with rich beetle-green sheen. Undercolour black, as dark as possible.

Plumage, Blue Variety, Male and Female: Any shade from light to dark blue, medium preferred, uniform in colour throughout, forming a solid clear blue free from mealiness.

Plumage, Buff Variety, Male and Female: Clear sound even buff throughout to skin, with greater lustre on male hackle, saddle and wing bow. No particular shade of buff standardised, but preference given to the paler lemon-buff shades. Even colour is more important than exact shade.

Plumage, Columbian Variety, Male: Pearl-white with black markings. Head and neck hackle white, with dense black stripe down middle of each feather, free from black edging and black tips.

Wyandotte
Partridge Male
White Female

Saddle white. Tail feathers and tail coverts glossy green-black, the coverts laced or not with white. Primaries black, or black edged with white. Secondaries black on inner edge, white outer. Remainder of plumage entirely white, of pearl-white shade, free from ticking. Undercolour either slate, blue-white or white.

Plumage, Female: Head white. Neck hackle intense black entirely surrounded with a white marginal fringe. Tail feathers black, the top pair permissibly laced with black. Primaries black, edged with white. Secondaries black inner edge, white outer. Remainder pearl-white, entirely free from ticking. Undercolour slate, blue-white or white.

Plumage, Gold Laced Variety, Male: Head rich golden bay, neck hackle similar with distinct black stripe down the middle of each feather, free from black tips or black edging. Saddle hackles to match neck. Back rich bay, free from black, not maroon coloured. Wing bow rich golden bay. Secondaries black on inner web and with wide gold stripe on outer web, the edge laced with black. Primaries black inner web and widely laced with gold on outer edge. Breast and underparts golden bay with well defined jet black lacing, free from double lacing or outer gold fringe. Lacing regular from throat to back of thighs, showing rich green lustre. Thighs and fluff black or dark slate powdered with gold, with clear lacing round hocks and outer side of thigh feathering.

Plumage, Female: Head rich golden bay. Neck hackle rich golden bay with distinct black stripe down middle of each feather, free from black edging or black tipping. Breast and back golden bay with well-defined jet black lacing free from double lacing or outer gold fringe, and with green lustre. Undercolour dark slate. Secondaries and primaries as in the male. Tail black with green sheen, coverts black with centres golden bay. Thighs and fluff black or dark slate powdered with gold.

Note: In both sexes brightness and uniformity of ground colour are of more importance than any particular shade.

Plumage, Mottled Variety, Male and Female: Rich black throughout with beetle-green sheen, the feathers tipped as evenly as possible with white.

Plumage, Partridge Variety, Male: This follows the usual black-red markings, with special attention to solid black striping in hackles, and orange neck and saddle hackles shading off to pale bright lemon colour at ends. Hackles free from black tipping, black fringes or light shaft, and striping should not run through at tips. Breast sound glossy black, breast and fluff free from red or grey ticking;

292

all undercolour black or dark grey, free from white. Back and shoulders rich bright scarlet, free of maroon or purple. Wing bar glossy black; primaries solid black, free from white or grizzle; secondaries rich bay outer web, black inner web and end—bay only to show when wing is closed. Tail, sickles and coverts black with rich sheen, free from white at roots. (*Note:* In catering for the modern demand for pale lemon fringes to hackles, density of black is reduced in striping and breast, undercolour usually becomes paler, and white fluff shows at junction of saddle and tail.)

Plumage, Female: Head and neck hackle of rich golden-yellow, the lower feathers finely pencilled; remainder of plumage to have sharply defined concentric black pencilling on a soft partridge-brown ground colour (often described as the colour of a dead oak leaf); red or yellow tinges objectionable. Pencilling to be fine and even, with three or more distinct rings, and to be clearly defined all over the body, especially in adults. Shaftiness is objectionable, especially on breast; and markings should continue well down the fluff, which should match feather in colour. Fluff brown, of same shade as ground colour. Thumb-nail markings are objectionable. Secondaries brown pencilled with black on outer web, black on inner web; pencilling to show when wing is closed. Tail black with brown markings and with clearly pencilled outer feathers.

Plumage, Silver Laced Variety, Male and Female: As described for Gold Laced, except that the ground is silver-white, free from yellow or straw tinge, instead of rich golden bay. Regularity of markings and quality of colour in all cases to count above any particular breadth of lacing.

Plumage, Silver Pencilled Variety, Male: Head silvery white, neck silvery white with glossy black stripe down the middle of each feather. Back silvery white, free from rust or straw colour. Saddle hackle to match neck. Wing bows silvery white, wing bar glossy green-black. Secondary wing feathers, outer web partly white, forming wing bay when closed, remainder black. Primaries black on inner and silver on outer webs. Breast black, free from ticking or lacing. Fluff solid black. Tail and sickles glossy green-black.

Plumage, Female: Head and neck hackles silvery white, striped with black. Back, breast and wings of steel grey ground colour, each feather plentifully pencilled with darker colour, in concentric bands following shape of feather. Bands as numerous as possible and fine. Fluff to match breast with as much pencilling as possible. Tail feathers black shading to grey. Tail coverts pencilled.

Plumage, White Variety, Male and Female: Pure white throughout, without straw tinge.

Other colours include White Laced Buffs, Violet Laced, Blue Laced, Buffs, and Cuckoos.

In both sexes: Beak bright yellow, except in marked and laced varieties, in which it may be horn, shaded with yellow. (*Note:* Yellow beaks are unobtainable in black males with dark under-colour, and beak colour in these should be black, shaded with yellow.) Eyes bright bay in all colours. Comb, face, wattles and ear-lobes bright red. Legs and feet bright yellow.

Weights: Male 24 to 28 oz.   Female 20 to 24 oz.

SCALE OF POINTS

White Variety

COLOUR	25
TYPE	25
HEAD	5
COMB	10
LOBES	5
EYES	5
LEG COLOUR	5
SIZE	10
CONDITION	10
	100

Black Variety

COLOUR	20
UNDERCOLOUR	15
TYPE	20
HEAD	5
COMB AND LOBES	10
LEG COLOUR	10
SIZE	12
CONDITION	8
	100

Partridge and Silver Pencilled Varieties, Male

HACKLE COLOUR	8
STRIPING	8
TOP COLOUR	8
BREAST	7
FLIGHTS	5

Continued on next page

Wyandotte Scale of Points—continued

TAIL	4
UNDERCOLOUR	4
FLUFF	4
COMB	7
EYES	5
LOBES AND WATTLES	4
TYPE	22
LEGS AND FEET	6
SIZE AND CONDITION	8
	100

Partridge and Silver Pencilled Varieties, Female

GROUND COLOUR	13
FORM OF PENCILLING	11
CLEARNESS OF PENCILLING	10
FLUFF	5
HACKLE	4
COMB	6
EYES	5
LOBES	4
TYPE	22
LEGS AND FEET	10
SIZE AND CONDITION	10
	100

Gold or Silver Laced Varieties

COMB AND HEAD	10
LOBES AND WATTLES	4
NECK	8
BREAST AND THIGHS	12
BACK	12
TAIL	6
WINGS	12
FLUFF	5
LEGS	5
SIZE AND CONDITION	14
SHAPE	12
	100

Columbian Variety

COMB	10

Continued on next page

Wyandotte Scale of Points—continued

EYES	5
LOBES AND WATTLES	5
HACKLE AND TAIL	15
BODY COLOUR	15
LEGS	5
TYPE AND SYMMETRY	35
CONDITION	10
	100

Serious Defects: Feathers on shanks or toes. Permanent white or yellow in ear-lobes, covering more than one-third of the surface. Comb other than rose or flopping or obstructing the sight. Shanks other than yellow. Any deformity. Slipped wings (which should be penalised strongly). Eyes not matching or other than bright bay. Conspicuous peppering on ground colour of laced varieties. Any form of double lacing in laced varieties.

OTHER BREEDS

Booted Bantams: These were once very popular in this country. They are similar to Belgian Barbus d'Uccle, but non-bearded. Now seldom seen. Weights slightly higher than d'Uccle. Known in Belgium as Sabelpoot.

Campines: Recent attempts to bantamise Campines are afoot, and at the 1969 International Show, some passable Gold Campine bantams appeared. They would probably have more future in bantamised form, if perfected and distributed. Suggested weights as for Hamburghs.

Croad Langshans: In the early nineteen-sixties a class was provided for Croad Langshan bantams, at first with a few entries, but latterly without support. Perseverance and attainment of the dark brown egg could make them a welcome permanent addition to the breeds of bantams.

Dorkings: Seldom seen nowadays, but are still bred in Scotland. It would be good to see them redistributed.

Nankins: They were once among the most widespread of all bantams, and are believed to be the progenitors of nearly all buff varieties of bantam. The male is a warm shade of buff, with red or reddish brown on the back and wings. The female is buff throughout, with the exception of the tail which has a considerable amount of black in it. Comb (rose) face, wattles and lobes bright red. Type somewhat resembling the Rosecomb, with full hackle, prominent breast, fairly short in back, with tail well curved. Beak and legs slate blue.

Rumpless: Tailless bantams, usually of Game character. They occasionally have classes at shows, appearing in any of the familiar Old English Game colours.

Sumatra Game: Attempted replicas of the large Sumatra Game appear occasionally at shows, and if really established could be most attractive.

Yokohamas. A few specimens of single combed Yokohamas, usually Duckwings, and usually far too large, have appeared at shows in recent years. Seen regularly on the Continent, in small numbers.

Other breeds of bantams have appeared briefly and seemingly vanished, such as Aseels, Crève-Cœurs, Redcaps, Scots Dumpies and Spanish. Should they reappear, the existing Large Fowl standards are there as a guide.

TURKEYS

I⊤ is strange that the first English Book of Standards of 1865 incorporated the turkey, without defining colours or varieties. Yet at the first English Show of 1845 the classification was for Whites and Any Other Colour. The first turkeys to reach England came from Spain about 1524. A Spanish explorer had discovered turkeys on the coast of Cumana, north of Venezuela in 1499, from which place specimens were shipped to Spain in 1500, and some birds reached France in 1516. It has been said also that turkeys reached England from Spain in 1821. Commercially these standards are now obsolete having been superseded by broad-breasted varieties.

BELTSVILLE WHITE

T⊢⊣ıs is a comparative newcomer to turkey breeds, and was developed in America to meet the consumer demand for a smaller type of bird. It rapidly achieved popularity in the States, and although adopted by British breeders has now given way to the Small White.

GENERAL CHARACTERISTICS: MALE AND FEMALE

Type: Body very symmetrical, medium length, and broad. Back broad throughout and flat. Breast medium length, broad, and slightly curved. Wings strong and neatly folded. Tail long and straight, carried in line with the back.

Head: Small length and breadth, and carunculated. Beak strong, curved, and well set. Eyes prominent, bright, and clear. Throat wattle medium in length and of fine texture, conforming to the size of the bird.

Neck: Of medium length in proportion to rest of the body.

Legs and Feet: Thighs of medium length and strong. Fluff short

298

and tight.   Shanks fine but strong, flat sided, fairly short.   Toes, four, straight and strong.

Plumage: Very tightly feathered.

COLOUR

Plumage of both sexes pure white throughout, free from creaminess, off-white or other tinge; black beard.   Beak light pinkish horn. Eyes, brown iris, black pupil.   Face, wattle, and caruncles bright rich red, but changeable to blue and white in the male.   Shanks and feet pink and flesh colour.   Toes pale horn.

Standard Weights: Cock 18 to 22 lb; cockerel 12 to 17 lb. Hen 10 to 12 lb; pullet 8 to 10 lb.   (Weights are minimum and maximum respectively in each case.)

SCALE OF POINTS

Defects in	Deduct
CONDITION	10
HEAD, NECK, AND WATTLE	10
COLOUR	10
SHAPE AND TYPE	55
LEGS AND FEET	15
	100

Serious Defects: Crooked or other deformity of the breast bone. Deep breasts with a pronounced knob on point of breast bone. Wry tail.   Feathers other than white.   Pronounced debeaking.

Disqualification: Any birds exceeding the weights laid down in the Standard.

BLUE

THIS breed has been developed on the lines of the American Slate turkey.   The feathers of the Slate are ashy-blue and may be dotted with black in any part of the plumage, whereas the Blue Standard calls for an even colour.

GENERAL CHARACTERISTICS: MALE AND FEMALE

These are the same as for the Buff turkey.

COLOUR

Plumage of both sexes a light or dark shade of sound and even blue, free from black or brown feathers. Beak, legs and feet slate-blue. Eyes dark to black. Face, jaws, wattle and caruncles bright rich red.

Standard Weights: Cockerels 18 to 25 lb; pullets 14 to 18 lb.

SCALE OF POINTS

Defects in	Deduct up to
SHAPE (TYPE)	25
HEAD, NECK AND WATTLE	15
LEGS AND FEET	10
WEIGHT	10
COLOUR	35
CONDITION	5
	100

Serious Defects: Presence of black or brown feathers. Crooked breast, wry tail or any other deformity. Pronounced debeaking.

BRITISH WHITE

WHITE turkeys were bred here in Europe from the early times, and the breed has been known also as the White Holland, and even the Austrian. It retains its name of White Holland in America, being Standardised there as such, while there, too, producers have now developed a small Beltsville White. To distinguish our own developed strains of the White variety, it is to be called the British White, as bred to our Standard.

GENERAL CHARACTERISTICS: MALE AND FEMALE

Type: Body long, deep and well rounded. Back curving with good slope to tail. Breast broad, full, long and straight. Wings strong and large. Tail long in proportion to body.

Head: Long, broad and carunculated. Beak strong, curved and well set. Eyes bright, bold and clear. Throat wattle large and pendent.

Neck: Long, curving backward towards tail.

White Turkey
Male

Legs and Feet: Thighs long and stout. Fluff short. Shanks large, strong, well rounded and of medium length. Toes, four, straight and strong and well spread.

COLOUR

Plumage of both sexes pure white with black tassel. Beak white to pale horn. Eyes, iris dark hazel, pupil blue-black. Face, wattle and caruncles, bright rich red, but changeable in the male to blue and white. Shanks and feet pink flesh. Toe-nails white to pale horn.

Standard Weights: Cock 28 lb upward, but not to exceed 38 lb; cockerel 18 to 28 lb. Hen 16 to 22 lb; pullet 14 to 18 lb.

SCALE OF POINTS

Defects in	Deduct up to
SHAPE (TYPE)	35
HEAD, NECK AND WATTLE	10
LEGS AND FEET	5
WEIGHT	25
COLOUR	15
CONDITION	10
	100

Serious Defects: Feathers other than white. Crooked breast. Wry tail. Any other deformity. Pronounced debeaking.

BRONZE

THE Bronze Turkey was first developed here as the Cambridge, the Norfolk Black being used in the crossings, and it became known as the Cambridge Bronze. Later importations of American Bronze turkeys resulted in crossings with the Cambridge Bronze to obtain greater size, until the name American Mammoth Bronze became generally known. It is now Standardised under the name Bronze.

GENERAL CHARACTERISTICS: MALE AND FEMALE

Type: Body long, deep and well rounded. Back curving with a good slope to the tail. Breast broad, full, long and straight. Wings strong and large. Tail long in proportion.

302

Head: Long, broad and carunculated. Beak strong and curved and well set. Eyes bright, bold and clear. Throat wattle large and pendent.

Neck: Long, and curving backwards towards the tail.

Legs and Feet: Thighs medium and stout. Fluff short. Shanks large, fairly long and strong. Toes, four, straight, strong and well spread.

COLOUR

Plumage of both sexes a good metallic bronze throughout, the female to have a slight ticking on the breast. Flights black with a definite white barring. Tail black and brown, with a good broad black band edged with white. Beak horn. Eyes, iris a dark hazel, pupil blue-black. Face, jaws, wattle and caruncles bright rich red. Shanks and toes black or horn. Toe-nails horn.

Standard Weights: Cock 30 to 40 lb; cockerel 25 to 35 lb. Hen 18 to 26 lb; pullet 14 to 22 lb.

SCALE OF POINTS

Defects in	Deduct up to
SHAPE (TYPE)	25
HEAD, NECK AND WATTLE	15
LEGS AND FEET	10
WEIGHT	25
COLOUR	20
CONDITION	5
	100

Serious Defects: Crooked breast, wry tail and any other deformity. Pronounced debeaking.

BUFF

IN their day our breeders developed excellent strains of the Buff turkey, which were often reckoned to be one of the best layers among the various breeds. It was not aptly named as the colour is actually a deep cinnamon-brown matching the colour of a dead beech leaf.

GENERAL CHARACTERISTICS: MALE AND FEMALE

These are the same as those of the British White, with the exception that the shanks are large, fairly long and strong.

COLOUR

Plumage of both sexes is a deep cinnamon-brown. Flights and secondaries white. Tail deep cinnamon-brown edged with white. Beak light horn. Eyes, iris dark hazel, pupil blue-black. Face, jaws, wattle and caruncles bright rich red. Shanks and toes pink and flesh. Toe-nails light horn.

Standard Weights: Cock 22 to 28 lb; cockerel 16 to 23 lb. Hen 12 to 18 lb; pullet 8 to 14 lb.

SCALE OF POINTS

Defects in	Deduct up to
SHAPE (TYPE)	20
HEAD, NECK AND WATTLE	15
LEGS AND FEET	10
WEIGHT	25
COLOUR	25
CONDITION	5
	100

Serious Defects: White in tail except in edging. Crooked breast, wry tail, or any other deformity. Pronounced debeaking.

NORFOLK BLACK

THE original turkey importations into this country were darks and, no doubt, the first variety to be developed here was the Norfolk Black in the county from which it derived its name. Specimens were shipped to America and the breed may well be claimed as the first of the domestic varieties of turkey.

GENERAL CHARACTERISTICS: MALE AND FEMALE

Type: Body fairly long and deep, particularly broad across the shoulders. Back broad and flat between the shoulders. Breast

Blue and Norfolk
Black Turkey
Males

not too long, well rounded, muscular and fleshy. Wings carried lightly. Tail long in proportion to body.

Head: Fairly long and broad, and carunculated. Beak strong curved and well set. Eyes bright, bold and clear. Throat wattle large and pendent.

Neck: Of medium length curving slightly backward with an alert carriage.

Legs and Feet: Legs short to medium length and set well apart. Thighs full and thick. Toes, four, straight, strong and well spread.

COLOUR

Plumage of both sexes a dense black. Beak, legs and feet black. Eyes dark to black. Face, jaws, wattle and caruncles bright rich red; but short black feathers on head and face not a fault.

Standard Weights: Cock 25 lb; cockerel 18 to 22 lb. Hen 13 to 15 lb; pullet 11 to 13 lb. Dead weight, not drawn: cockerel 20 lb (max.); pullet 12 lb (max.).

SCALE OF POINTS

Defects in	Deduct up to
SHAPE (TYPE)	20
HEAD, NECK AND WATTLE	15
LEGS AND FEET	10
WEIGHT	25
COLOUR	25
CONDITION	5
	100

Serious Defects: Bronze in undercolour or wings. White flecks on thigh or wing feathers. Crooked breast, wry tail, or any other deformity. Pronounced debeaking.

Other Breeds: Among other breeds of turkey occasionally exhibited in this country may be mentioned the Lavender or Slate, slate or ash-blue dotted (but not laced or spangled) with black; and the Italian grey-black, the feathers edged with a clear grey.

DUCKS

IT is generally accepted that all breeds of ducks, with the exception of the Muscovy, originated from the wild Mallard. This is quite clear with a breed like the Rouen, and some consider that the Black East Indian and the Cayuga originated from sports of the Mallard. It is possible to understand, too, the original white ducks of this country coming as Mallard sports. They may have been developed for body size and table qualities, by domestication and selection, resulting eventually in the Aylesbury as we know it today. In the make-up of the Khaki Campbell the wild Mallard also played its part.

The British Waterfowl Association classifies the following as Ornamental Ducks: White Decoy, Brown Decoy, Silver Appleyard Bantams, Black East Indian, Carolina, Mandarin and all other British or foreign breeds of wild duck.

AYLESBURY

ORIGIN: British

THE Aylesbury derives its name from the town of Aylesbury in Buckinghamshire. At the first poultry show of 1845, a class was provided for Aylesbury or Other White Variety, and another for Any Other Variety. No doubt there were white ducks in this country for centuries before, and from them was developed by judicious selection for table purposes the White Aylesbury, which is today Britain's table breed de luxe. Once Standardised as a breed it was developed by selecting for its distinctive characteristics, which separated it from all other white breeds of ducks.

GENERAL CHARACTERISTICS: MALE AND FEMALE

Carriage: Horizontal, the keel practically parallel with the ground.

Type: Body long, broad and very deep, showing a good keel. Back straight, almost flat. Breast full and prominent. Keel quite straight from breast to stern. Wings strong and carried closely to the sides, fairly high but not touching across the saddle. Tail short, only slightly elevated, and composed of stiff feathers, the drake's having two or three well-curled feathers in the centre.

Head: Strong and powerful, with eyes as near the top of the skull as possible. Bill strong and wedge-shaped. When viewed from the side the outline is almost straight from the top of the skull, the head and bill measuring from six to eight inches. Eyes full.

Neck: Curved and strong.

Legs and Feet: Legs very strong and short, the bones thick, set to balance a level carriage. Feet straight and webbed.

Plumage: Bright and glossy, resembling satin.

COLOUR

Plumage of both sexes white. Bill pink-white or flesh. Eyes dark. Legs and webs bright orange.

Standard Weights: Drake 10 lb. Duck 9 lb.

SCALE OF POINTS

TYPE	10
SIZE	20
HEAD AND BILL	20
EYES	8
KEEL	10
COLOUR	10
NECK	5
LEGS AND FEET	5
CONDITION	12
	100

Serious Defects: Plumage other than white. Bill other than white or flesh-pink. Heavy behind. Any deformity.

Aylesbury
Drake and Duck

BLACK EAST INDIAN

ORIGIN: American

THE Black East Indian is described in the first Book of Standards of 1865, but it has had other names such as Buenos Ayres, Labrador, and Black Brazilian. With weights of $1\frac{1}{2}$ to $1\frac{3}{4}$ lb for the duck and of 2 lb for the drake, it might be regarded as the " bantam " of the duck breeds. Many consider this a black sport from the Mallard.

GENERAL CHARACTERISTICS: MALE AND FEMALE

Carriage: Lively, smart, symmetrical and clear of the ground from breast to stern.

Type: Body short and broad. Breast round and prominent.

Head: Neat and round, with high skull. Bill short and fairly broad, well set in a straight line from the tip of the eye. Eyes full.

Neck: Short.

Legs and Feet: Legs of medium length, placed midway in the body. Feet straight and webbed.

Plumage: Bright and glossy.

COLOUR

Plumage of both sexes a very lustrous intense beetle-green black, free from purple or white feathers, a brown or purple tinge being objectionable but not a disqualification. Bill slate-black, olive patches not objectionable. Eyes dark. Legs and webs as black as possible, but becoming more or less orange with age.

Standard Weights: Drake 2 lb. Duck $1\frac{1}{2}$ to $1\frac{3}{4}$ lb.

SCALE OF POINTS

TYPE	20
SIZE	20
HEAD, BILL AND NECK	15
LEGS AND FEET	5
COLOUR	30
CONDITION	10
	100

Serious Defects: White and purple feathers. Slipped wing. Dished bill. Any deformity.

CAMPBELL

ORIGIN: British

I T was the wild Mallard that played its part in the make-up of the Khaki Campbell, together with blood of the Fawn-and-White Runner and that of the Rouen. Introduced in 1901 by its originator, Mrs. Campbell of Uley, Glos., it was her special desire to keep the breed for prolific egg-laying, so that a very elementary Standard was at first publicised. In this way the high egg-producing properties of the breed were maintained. The White Campbell came as a sport from the Khaki. The Dark Campbell was created by Mr. H. R. S. Humphreys in Devon, to make sex-linkage in ducks possible.

GENERAL CHARACTERISTICS: MALE AND FEMALE

Carriage: Alert, slightly upright and symmetrical, the head carried high, with shoulders higher than the saddle, and the back showing a gentle slant from shoulder to saddle; the whole carriage not too erect but not as low as to cause waddling. Activity and foraging power to be retained without loss of depth and width of body generally.

Type: Body deep, wide and compact, appearing slightly compressed, retaining depth throughout, especially from shoulders to chest and from middle of back through to thighs; broad and well-rounded front. Back wide, flat and of medium length, gently sloping with shoulders higher than saddle. Abdomen well developed at rear of legs, but not sagging; well-rounded underline of breast and stern. Wings closely carried and rather high. Tail short and small, rising slightly, the drake's with the usual curled feathers.

Head: Refined in jaw and skull. Face full and smooth. Bill proportionate, of medium length, depth and width, well set in a straight line with the top of the skull. Eyes full, bold and bright, showing alertness and expression, high in skull and prominent.

Neck: Of medium length, slender and refined, almost erect.

Legs and Feet: Legs of medium length, and well apart to allow of good abdominal development; not too far back. Feet straight and webbed.

Plumage: Tight and silky, giving a sleek appearance.

Quality and Refinement: While aiming at good body size emphasis should be placed upon quality or refinement in general, i.e. neat bone, sleek silky plumage, smooth face, fine head points, etc., with absence of coarseness and sluggishness.

COLOUR

Dark Variety: Drake's plumage: Head and neck beetle-green. Shoulders, breast, underparts and flank light brown, each feather finely pencilled with dark grey-brown, gradually shading to grey at stern close up to the vent, followed by beetle-green feathers with purplish tinge up to the tail coverts. Tail feathers dark grey-brown; coverts beetle-green with purplish tinge or reflection, also curled feathers in centre. Wing bow dark grey-brown laced with light brown; bar broad purplish green band, edged with a thin light grey line on each side; flights and secondaries dark brown. Bill bluish-green with black bean-shaped mark at the tip. Eyes brown. Legs and feet bright orange.

Duck's plumage: Head and neck dark brown. Shoulders, breast and flank light brown, each feather broadly pencilled with dark brown, becoming brown towards stern, with lighter outer lacing, followed by beetle-green feathers at the rump. Back and wing bow dark brown, outer laced with lighter brown. Wing bar as in drake, but less lustrous. Tail and wing feathers dark brown. Bill slaty-brown with a black bean-shaped mark at the tip. Eyes brown. Legs and feet as near body colour as possible.

Khaki Variety: Drake's plumage: Head, neck, stern and wing bar green-bronze. Remainder of plumage an even shade of warm khaki shading off to lighter khaki towards the lower parts of the breast. Bill greenish-blue, the darker the better. Legs and webs dark orange.

Duck's plumage an even shade of warm khaki. Head and neck a slightly darker shade of khaki. Breast lightly laced on close examination. Back laced. Wings khaki, sound top and under. Bill greenish to slaty-black. Legs and webs as near the body colour as possible.

White Variety: Plumage of both sexes pure white throughout. Bill, legs and webs orange. Eyes grey-blue.

Standard Weights: Drake 5 to 5½ lb. Duck in laying condition 4½ to 5 lb.

Campbell
Khaki Drake
and Duck

SCALE OF POINTS

TYPE (SHAPE AND CARRIAGE)................ 25
SIZE AND SYMMETRY........................ 10
HEAD POINTS.................................. 10
LEGS AND FEET.............................. 5
COLOUR...................................... 25
QUALITY AND REFINEMENT.................. 15
CONDITION 10
                                                 100

Serious Defects: Dark: Yellow bill. White bib or white neck ring. Any deformity. Green eggs. Coarseness. **Khaki:** Yellow bill. White bib or white neck ring. Streak from eyes in duck. White or light underpart or top of wings in either sex. White in wing bar. Any deformity. Green eggs. Anything that interferes with production. **White:** Excessive weight or coarseness. Flesh-coloured bill. Any deformity.

CAYUGA

ORIGIN: American

Many consider that the Cayuga was bred from the Black East Indian along larger lines. The breed takes its name from Lake Cayuga, New York. It was in 1851 that black ducks made their appearance on the lake, and specimens were sent to this country.

GENERAL CHARACTERISTICS: MALE AND FEMALE

Carriage: Lively, clear of the ground from breast to stern.

Type: Body long, broad and deep. Breast prominent, keel well forward and forming a straight underline from stem to stern. Tail carried well out and closely folded, the drake's having two or three well-curled feathers in the centre.

Head: Large. Bill long, wide and flat, well set in a straight line from the tip of the eye. Eyes full.

Neck: Long and tapering with a graceful curve.

Legs and Feet: Legs large, strong boned, placed midway in the body, giving the bird a carriage similar to that of the Rouen. Feet straight and webbed.

COLOUR

Plumage of both sexes a very lustrous green-black, free from purple or white, the whole of the back and upper part of wings, the breast, and underparts of body deep black, the wings naturally more lustrous than the rest of the body plumage; a brown or purple tinge is objectionable, although not a disqualification. Bill slate-black, with dense black saddle in the centre, but not touching the sides or coming within an inch of the end, the bean black. Eyes black. Legs and webs dull orange-brown.

Standard Weights: Drake 8 lb. Duck 7 lb.

SCALE OF POINTS

TYPE	30
SIZE	20
NECK	5
TAIL	5
HEAD AND BILL	10
CONDITION	10
LEGS AND FEET	5
COLOUR	15
	100

Serious Defects: Red or white feathers. Orange-coloured bill. Dished bill. Any deformity.

CRESTED

ORIGIN: British

THE Crested is one of the less common breeds of duck, somewhat like the Orpington in size and type. In breeding Crested ducks not all the ducklings have crests, the plain being distinguishable from the crested at birth. Their utility qualities are comparable with those of the Orpington.

White Crested
Duck

GENERAL CHARACTERISTICS: MALE AND FEMALE

Carriage: Somewhat erect.
Head: Long and straight. Crest globular, large, set evenly on the skull. Bill long and broad. Eyes large and bright.
Neck: Rather long, slightly arched.
Body: Long, broad and fairly deep, full round breast, long broad back. Strong wings, carried closely. Short tail, similar to that of the Aylesbury.
Legs: Short and strong. Toes straight and connected by web.

COLOUR

Any colour is permitted.

Standard Weights: Drake 7 lb. Duck 6 lb.

SCALE OF POINTS

CREST	25
TYPE	25
SIZE	15
HEAD AND BILL	10
CONDITION	15
NECK	5
LEGS AND FEET	5
	100

Serious Defects: Slipped wings. Any deformity.

DECOY

ORIGIN: British

THESE birds have been standardised since the first edition of this book. There are two colours, white and brown, and emphasis is placed upon alertness in looks and movement.

GENERAL CHARACTERISTICS: MALE AND FEMALE

Carriage: Carried well, and nearly level from breast to stern.

Type: Body very small and compact, broad and deep.

Head: Small, neat and round, with high skull. Bill short and broad, set deep into the skull. Eyes round, full and alert.

Neck: Short.

Legs and Feet: Legs short, set midway in the body. Feet straight and webbed.

COLOUR

White Variety: Plumage of both sexes pure white all over, free from sappy yellow colour. Bill bright orange-yellow. Eyes dark. Legs and feet orange-yellow.

Brown Variety: Plumage of both sexes resembles the same colour as the Mallard. Drake's bill an olive-green, the duck's as dark as possible. Eyes dark. Legs and feet a dark brown, muddy colour.

Standard Weights: Drake 20 to 24 oz. Duck 16 to 20 oz.

SCALE OF POINTS

TYPE	30
SIZE	20
COLOUR	20
HEAD, NECK AND BILL	15
LEGS AND FEET	5
CONDITION	10
	100

Note: Decoys, especially the White, should be noticeably alert in looks and movement. Thin bodied, boat-shaped bodies, thin bills and flat skulls must be avoided.

INDIAN RUNNER

ORIGIN: Asiatic

THE era of the high egg-laying breeds of ducks started with the introduction of the Indian Runner into this country from Malaya. A ship's captain brought home Fawns, Fawn-and-Whites, and Whites, distributing them among his friends in Dumfries-shire and Cumberland. They proved prolific layers, and there was a class of Fawn Runners at the Dumfries show in 1876 but the Fawn-and-

Whites were not exhibited until 1896. The Indian Runner Duck club's Standard of 1907 described only the Fawn-and-White, that of 1913 recognised also the Fawn, while the 1926 Standard described the Black and the Chocolate varieties.

GENERAL CHARACTERISTICS: MALE AND FEMALE

Carriage: Upright and active. The " angle " of inclination of the body to the horizontal varies from 50 to 80 degrees when the bird is on the move and not alarmed; but when standing at attention, or excited, or specially trained for the show pen, it may assume an almost perpendicular pose.

Type: Body slim, elongated and rounded, but slightly flattened across the shoulders. At the lower extremity the front line sweeps gradually round to the tail, which is neat and compact and almost in a line with the body or horizontally, but in some excellent birds slightly elevated or tilted upwards—the position of the tail varying with the attitude of the duck; however habitually upturned sterns and tails (as in the Pekin duck) are considered objectionable. Stern short compared with other breeds, the prominence of the abdomen and stern varying in ducks according to the season and the age of the bird, being fuller when in lay; but a large pendulous abdomen and long stern, or a " cut-away " abdomen and stern in young ducks should be avoided. Wings small in proportion to the size of the bird, tightly packed to the body and well tucked up, the tips of the long flights of the opposite wings crossing each other over the rump, more particularly when the bird is standing at attention. At the upper extremity the body gradually and imperceptibly contracts to form a funnel-shaped process, which again, without obvious junction merges into the neck proper, the lower or thickest portion of this funnel-shaped process or " neck expansion " being reckoned as part of the body.

Note: Total length of drake 26 to 32 in, and duck 24 to 28 in. Length of neck proper, from top of skull to where it joins the thick part of the " funnel " about one-third the total length of the bird, not less. Measurements should be taken with the bird fully extended in a straight line, the bill and head in a line with the neck and body, and the legs and feet in the same straight line, the measurements being from the tip of the bill to the tip of the middle toe.

Head: Lean and racy looking, and with the bill wedge-shaped. Skull flat on top, and the eye-socket set so high that its upper margin seems almost to project above the line of the skull. Eyes

319

full, bright, very alert and intelligent. Bill strong and deep at the base where it fits imperceptibly into the skull. There should be no indication of a joint or " stop ". The upper mandible should be very strong and nicely ridged from side to side, and the line of the lower mandible straight also, with no depression or hollow in the upper line from its tip to its base. The outline should run with a clean sweep from the tip of the bill to the back of the skull. The length and depth of the bill varies, but should never be out of balance or harmony with the rest of the head and the lines of the bird as a whole.

Neck: Long and slender, and when the bird is on the move or standing at attention, almost in a line with the body, the head being high and slightly forward. The thinnest part is approximately where, in Fawn drakes, the dark bronze of the head and upper neck joins the lower or fawn of the neck proper. The muscular part should be well marked, rounded and stand out from the windpipe and gullet, the extreme hardness of feather helping to accentuate this. The neck should be neatly fitted to the head.

Legs and Feet: Legs set far back to allow of upright carriage. Thighs strong and muscular, longer than in most breeds. Shanks short and feet supple and webbed. There should be sufficient width between the legs to allow of free egg production, but not as much as to cause the duck, on actual test, to roll or waddle when in motion.

Plumage: Tight and hard.

COLOUR

Black Variety: Plumage of both sexes solid black with metallic lustre like the Black East Indian. There should be no grey under the chin or wings, no grey wing ribbons, and no " chain armour " on the breast. Bill black. Legs and webs black or very dark tan.

Chocolate Variety: Plumage a rich chocolate throughout, the drake, on assuming male plumage, darker than the duck, but the ground work is the same. Bill, legs and feet black.

Fawn Variety: Drake's plumage: Head and upper part of the neck dark bronze with metallic sheen, which may show a faint green tinge meeting the colour of the lower part of the neck with a clean cut or the lower colour merging into it imperceptibly. Lower neck and " neck expansion " rich brown-red continued on to breast, over the top of the shoulders and upwards to where it joins the head and upper neck colour, merging gradually on the back and breast into the body colour. Lower chest, flanks and

Indian Runner
White Drake,
Fawn Duck

abdomen French-grey, made up of very minute and dense peppering of dark brown, or almost black, dots on a nearly white ground, giving a general grey effect without any show of white, the grey extending beyond the vent until it meets the dark or almost black feathers of the cushion under the tail. Scapulars (the long pointed feathers on each side of the back covering the roots of the wings) red-brown, peppered. Back and rump deep brown, almost black. Tail (fan feathers and curl) dark brown, almost black. Wings: bow fawn, not pencilled; bar fawn corresponding with the coverts in the lower part, the upper part darker brown corresponding with the secondaries, which are black-brown with slight metallic lustre; primaries brown, fairly dark. When the drake is in " eclipse " or duck plumage he more closely approaches the duck in colour. All the dominant colours fade, but his head and neck are darker than the duck's; the body becomes a dirty fawn or ash, with perhaps some rustiness on the breast. Bill pure black to olive-green, mottled with black, and black bean. Legs and webs black, or dark tan, mottled with black.

Duck's plumage: The general plumage colour is an almost uniform warm ginger-fawn, with no marked variation of shade but having a slightly mottled or speckled appearance. When closely examined the head, neck, lower part of chest and abdomen may appear a shade lighter than the rest of the body. Each feather of the head and neck has a fine line of dark red-brown, giving a ticked appearance. Lower part of neck and " neck expansion " a shade warmer, each feather pencilled with a warm red-brown. Scapulars rich ginger-fawn, a shade darker than the shoulder and back, with well marked red-brown pencilling. Wing: bow a shade lighter than the scapulars but darkening towards the bar, the feathers pencilled as before; secondaries warm red-brown; primaries a shade lighter. Back and rump darker, the pencilling being richer and more marked, but the ground colour becomes lighter and warmer towards the tail. Tail lighter than upper parts of body, each feather pencilled. Belly lighter than upper parts of body, about the same shade of fawn as the head and neck, becoming a trifle darker on the tail-cushion, all feathers pencilled. In young ducks that have just completed the first adult moult there is often a rosy tinge on the lower neck-expansion, upper part of breast and shoulders, but this soon fades away. Some ducks have a cream or light coloured narrow band in the wing bar owing to the upper part of each feather being of a lighter or almost cream shade edged or laced again with the normal dark shade. Bill black. Eye iris golden-brown. Legs and webs black or dark tan.

Fawn-and-White Variety: Plumage of both sexes: The cap and cheek markings in the duck nearly the same shade of fawn as the body colour, but dull bronze-green in the drake; the cap separated from the cheek markings by a projection from the white of the neck extending up to, and in most cases terminating in, a narrow line more or less encircling the eye. The cap should come round the back of the skull with a clean sweep—there should be no " tails " to it. The cheek markings should not extend on to the neck. Bill divided from head markings by a narrow prolongation of the neck-white from $\frac{1}{8}$ to $\frac{1}{4}$ in wide, extending or projecting from the white underneath the chin. Neck pure white to about where the " neck-expansion " begins, and meeting the body with a clean cut. Body uniform soft warm or ginger-fawn to the skin, the undercolour not of a different shade. The rump and tail of the drake, including the under surface of his tail, a similar hue to his head. When closely examined the coloured body-feathers of the drake show a soft warm ground, slightly peppered with a rather warm shade, as only the outer edge of the feather is visible. The colour seems solid and more ruddy than that of the duck. The duck should have the same shade of fawn as the Fawn duck. The fawn and the white should meet on the breast with an even cut about half-way between the point of the breast bone and the legs. The base of the neck, upper part of wings, back and tail should be as nearly as possible the same colour as the fawn of the breast, and from the fawn of the back an irregular branch on either side extending downwards on the thighs to, or nearly to, the hough. The white of the breast extends downwards between the legs to beyond the vent and may overlap the thighs in part; if the bird is coloured between the ribs and thigh it is termed " foul flanked ". Wings, primaries, secondaries, and lower part of bow, pure white, which gives the appearance of a " heart " laid flat on the bird's back. Bill light orange-yellow in young birds; entirely, or almost entirely, dull cucumber in adult duck and green-yellow in the adult drake. Legs and feet orange-red.

White Variety: Plumage of both sexes a pure white throughout. Bill, legs and webs orange-yellow. Eyes, iris blue.

Standard Weights: Drake $3\frac{1}{2}$ to 5 lb. Duck 3 to $4\frac{1}{2}$ lb. Birds bred and shown in the same year as hatched may be accepted for competition at $\frac{1}{2}$ lb less.

SCALE OF POINTS

BODY—SHAPE AND GENERAL APPEARANCE OF,
INCLUDING LOWER PART OF NECK, LEGS
AND FEET..................................... 35
CARRIAGE AND ACTION...................... 30
HEAD, EYES, BILL AND NECK, EXCLUSIVE OF
LOWER NECK EXPANSION................... 20
COLOUR AND CONDITION..................... 15
                                          ——
                                          100

Serious Defects: Above and below standard weights and measurements. Body squat and short, oval or flattened. Domed skull with central position of eyes. Bill dished, weak, " Roman " under-curved or flat. Neck thick and short, swan or curved. Neck-expansion too far back on body causing a chesty appearance in front with a hollow behind. Legs set too far forward causing poor carriage. Waddling or rolling gait. Natural carriage in any duck below 40 degrees. Long stern. Wry tail. Flattened back. Slipped wing or any deformity. In Fawns, white anywhere; eyebrows or eye-stripes; light or cream wings (bows, coverts and flights); in the duck, blue or green wing bars; orange or yellow bill, feet or legs.

Note: In the Standard issued by the Indian Runner Duck Club of Great Britain appear the following remarks: " The Indian Runner, compared with the larger domesticated varieties, is a small, hard-feathered duck of very upright carriage, active habits, and moves with a straight-out walk or run, quite distinctive and different from the roll or waddle of the other domesticated ducks. It is a great traveller and forager, being much less dependent on the proximity of deep or swimming water than other ducks, and it is a prolific layer. Its body appears elongated and somewhat cylindrical, the legs being set on very far back. . . . It has long been an axiom of the Indian Runner Duck Club that a duck which cannot maintain a natural carriage of at least 40 degrees to the horizontal will not be considered a pure Runner, however good its other points may be. . . . The only varieties recognised by the Club are Whole Fawns, Whole Whites, Fawn-and-White, Blacks, Chocolates. The multiplication of varieties is not advisable as tending to create confusion."

MUSCOVY

ORIGIN: American

THE Muscovy has also been known as the Musk duck, and the Brazilian. It is a distinct species, and wild ancestors of it were found in South America, which must have credit for the original source. Under domestication, our breeders have greatly increased the size of the ducks, and also perfected their colour and markings. In more recent years the breed has become very popular, and specimens are seen at most shows where waterfowl are scheduled.

GENERAL CHARACTERISTICS: MALE AND FEMALE

Carriage: Low and jaunty.

Type: Body broad, deep, very long and powerful. Breast full, well rounded and carried low; keel long and well fleshed, just clear of ground, slightly rounded from stem to stern. Wings very strong, long and carried high. Tail long and carried low to give the body a longer appearance to the eye and a slightly curved outline to the top of the body.

Head: Large, and adorned with a small crest of feathers (more pronounced in the male than female) which are raised erect in excitement or alarm. Caruncles on face and over base of the bill. Bill wide, strong, of medium length and slightly curved. Eyes large, with wild or fierce expression.

Neck: Of medium length, strong and almost erect.

Legs and Feet: Legs strong, wide apart and fairly short. Thighs short, strong and well fleshed. Feet straight, webbed, with pronounced toe-nails.

Handling: Hard, well-fleshed, muscular.

Plumage: Close.

COLOUR

White-winged Black Variety: Plumage of both sexes a dense black throughout, except white wing bows. The black to carry a metallic green sheen, or lustre, with bronze on the breast and parts of the neck.

White-winged Blue Variety.   Plumage of both sexes blue except for white wing bows.

Black Variety: Plumage of both sexes a dense beetle-green black throughout, with bronze on the breast and parts of the neck.

White Variety: Plumage of both sexes pure white throughout.

Blue Variety: Plumage of both sexes light or dark shade of blue.

Black-and-White Variety: Plumage of both sexes black and white, with defined regularity of markings.

Blue-and-White Variety: Plumage of both sexes blue and white, with defined regularity of markings.

Variations in Colour: There are colour variations according to the countries of importation. In the Black-and-White, also the Blue-and-White, it is customary in some countries for the black or blue to predominate in winning specimens at the shows. Bill in all varieties yellow and black, red, flesh or lighter shade at point. Face and caruncles red or black. Eyes, from yellow and brown to blue.

Standard Weights: Drake 10 to 14 lb. Duck 5 to 7 lb. (It is a characteristic of the breed for the male to be about twice the size of the female.)

SCALE OF POINTS

SHAPE AND CARRIAGE	40
HEAD (INCLUDING CREST AND CARUNCULATIONS)	20
SIZE	20
CONDITION	10
COLOUR	10
	100

ORPINGTON

ORIGIN: British

IT was from the blending of the Indian Runner, Rouen and Aylesbury that Mr. W. Cook in Kent made the Buff Orpington, intending it to be a dual-purpose breed. Its introduction followed that of the Khaki Campbell, and it has been said that the originator was trying to make a strain of Khaki duck. At one time very popular for its high laying qualities combined with table properties and also its beauty of plumage and colouring, it has lost ground of late years.

GENERAL CHARACTERISTICS: MALE AND FEMALE

Carriage: Slightly elevated at the shoulders, i.e., not quite as horizontal as the Aylesbury, but avoiding any tendency to the upright carriage of the Pekin or Indian Runner.

Type: Body long and broad, deep, without any sign of keel. Back perfectly straight. Breast full and round. Wings strong and carried closely to the sides. Tail small, compact and rising slightly from the line of the back, the drake's having two or three curled feathers in the centre. When in lay the duck's abdomen should be nearly touching the ground.

Head: Fine and oval in shape, with narrow skull. Bill of moderate length, the upper mandible straight from bean to base in line with the highest point of the skull. Eyes large and bold, set high in the head. Deep set and scowling eyes are most objectionable.

Neck: Of moderate length, slender and upright.

Legs and Feet: Legs of moderate length, strong and set well apart. Feet straight and webbed.

Plumage: Tight and glossy.

COLOUR

Drake's plumage a rich even shade of deep red-buff throughout, free from lacing, barring and pencilling except that head and neck are seal-brown with bright gloss, but complete absence of beetle-green, the seal-brown to terminate in a sharply defined line all the way round the neck. The rump red-brown, as free as possible from " blue ".

Duck's plumage is similar to that of the drake's body, and free from blue, brown or white feathers.

In both sexes: Bill orange with dark bean. Eyes brown iris and blue pupil. Legs and webs bright orange-red.

Standard Weights: Drake not under 5 or over 7½ lb. Duck not under 5 or over 7 lb.

SCALE OF POINTS

TYPE AND SIZE	40
HEAD AND EYES	10
COLOUR	40
CONDITION	10
	100

Serious Defects: Colour other than stated. Twisted wings, wry tail, humped back or any other physical deformity. In the drake grey, silver, or blue head, white feathers in neck, brown secondaries, beetle-green on any part, very green bill. In the duck very heavy lacing, strong light line over the eyes, white feathers on neck or breast, brown feathers, green bill.

PEKIN

ORIGIN: Asiatic

Bred in China, the Pekin reached this country around 1874, and about the same time stock also went to America. English breeders called for a plumage of " buff canary, sound and uniform, or deep cream, the former preferred ". With judges showing preference for the " buff canary " plumage, the breed died out. In America, however, the Pekin became the producer of high-class table ducklings, with the Standard plumage of " creamy white ". Today there are a number of breeders of Pekins of this type, and a change in the new Standard in regard to plumage colouring may help to popularise the breed.

GENERAL CHARACTERISTICS: MALE AND FEMALE

Carriage: Almost upright, elevated in front and sloping downwards to rear.

Type: Body broad and of medium length and without any indication of keel except a little between the legs. Back broad. Breast broad and full followed in underline by the keel (which shows very slightly between the legs) to a broad, deep paunch and stern carried just clear of the ground. Wings short, carried closely to the sides. Tail well spread and carried high, the drake's having two or three curled feathers on top. A good description of the general shape of the Pekin is that it resembles a small wide boat standing almost on its stern, and the bow leaning slightly forward.

Head: Large and broad and round, with high skull, rising rather abruptly from the base of the bill, and heavy cheeks. Bill short, broad and thick, slightly convex but not dished. Eyes partly shaded by heavy eyebrows and bulky cheeks.

Neck: Long and thick, carried well forward in a graceful arch or curve and with slightly gulleted throat.

Legs and Feet: Legs strong and stout, set well back and causing erect carriage. Feet straight and webbed.

Plumage: Very abundant, thighs and fluff well furnished with long, soft downy feathers.

COLOUR

Plumage of both sexes cream or creamy white. Bill bright orange and free from black marks or spots. Eyes dark lead-blue. Legs and webs bright orange.

Standard Weights: Drake 9 lb. Duck 8 lb.

SCALE OF POINTS

TYPE	25
SIZE	20
HEAD (BILL 10, EYES 5)	15
NECK	5
COLOUR	15
TAIL	5
LEGS AND FEET	5
CONDITION	10
	100

Serious Defects: Black marks or spots on bill. Any deformity.

ROUEN

ORIGIN: French

CONFIRMATION of the Mallard as the progenitor of duck breeds is found in the Rouen, which so closely resembles it in plumage markings, while the Rouen drake moults into duck plumage in the summer like the wild Mallard drake. The Rouen undoubtedly came from Rouen in France, and was known also as the Rhone duck.

When first brought to England from France the breed was developed for table properties and was used in table crossings. Later it was bred, as today, for beauty of plumage and markings.

GENERAL CHARACTERISTICS: MALE AND FEMALE

Carriage: Carried horizontal, the keel parallel with ground and just clear of it.

Type: Body long, broad and square. Keel deep, just clear of the ground from stem to stern. Breast broad and deep. Wings large and well tucked to the sides. Tail very slightly elevated, the drake's having two or three curled feathers in the centre.

Head: Massive. Bill long, wide and flat, set on in a straight line from the tip of the eye. Eyes bold.

Neck: Long, tapering and erect, slightly curved but not arched.

Legs and Feet: Legs of medium length. Shanks stout and well set to balance the body in a straight line. Feet straight and webbed.

COLOUR

Drake's plumage: Head and neck rich iridescent green to within about an inch of the shoulders where the ring appears. Ring perfectly white and cleanly cut, dividing the neck and breast colours, but not quite encircling the neck, leaving a small space at the back. Breast rich claret, coming well under; cleanly cut, not running into the body colour, quite free from white pencilling or chain armour; chain armour or flank pencilling rich blue French-grey, well pencilled across with glossy black, perfectly free from white, rust or iron. Stern same as flank, very boldly pencilled close up to the vent, finishing in an indistinct curved line (perfectly free from white) followed by rich black feathers up to the tail coverts. Tail coverts black or slate-black with brown tinge, with two or three green-black curled feathers in the centre. Back and rump rich green-black from between the shoulders to the rump. Wings: large coverts pale clear grey; small coverts French-grey very finely pencilled; pinion coverts dark grey or slate-black; bars (two, composed of one line of white in the centre of the small coverts) grey tipped with black, also forming a line at the base of flight coverts, the latter feathers slate-black on the upper side of the quill and rich iridescent blue on the lower side, each of these feathers tipped with white at the end of the lower side, forming two distinct white bars (the pinion bar being edged with black) with a bold blue ribbon-mark between the two, each colour being clear and distinct and making a striking contrast; flights slate-black with brown tinge free from white. Markings throughout the whole plumage should be cleanly cut and well defined in every detail, the colours distinct and not shading into each other. Bill bright green-yellow, with black bean at the tip. Eyes dark hazel. Legs and webs bright brick red.

Duck's plumage: Head rich (golden, almond or chestnut) brown with a wide brown-black line from the base of the bill to the neck, and very bold, black lines across the head, above and below the eyes, filled in with smaller lines. Neck the same colour as the head, with a wide brown line at the back from the shoulders, shading to black at the head. Wings: bars, two distinct white bars with a bold blue ribbon-mark between, as in the drake; flights slate-black with brown tinge, no white. Remainder of plumage rich (golden almond or chestnut) brown of level shade, every feather distinctly pencilled from throat and breast to flank and stern, the markings to be rich black or very dark brown, the black pencilling on the rump having a green lustre. Bill bright orange, with black bean at tip, and black saddle extending almost to each side and about two-thirds down towards the tip. Eyes dark hazel. Legs and webs dull orange-brown.

Standard Weights: Drake 10 lb. Duck 9 lb.

SCALE OF POINTS

Drake

TYPE	10
SIZE	10
HEAD	5
LEGS AND FEET	5
COLOUR (BREAST 10, BILL 5, NECK 5, CHAIN ARMOUR 10, BACK AND RUMP 5, WINGS 5, STERN 5, TAIL 5)	50
MARKINGS	10
CONDITION	10
	100

Duck

TYPE	10
SIZE	10
HEAD	5
LEGS AND FEET	5
COLOUR (GROUND 15, BILL 10, HEAD 5, NECK 5, WINGS 5)	40
PENCILLING	20
CONDITION	10
	100

13*

Serious Defects: Leaden bill. No wing bars. White flights. Stern broken down. Wings down or twisted. Any deformity. In the drake, black saddle, black bill or minus ring (on neck); and in the duck, white or approaching white, ring (on neck).

OTHER BREEDS

In the list below are mentioned some breeds of ducks that are more or less moribund if not extinct, although such a statement must be made with reserve as there are probably specimens of some of these breeds still in existence.

Bali: Origin East Indian. Type: Head resembling that of a Runner, as also shape of body with a crest mainly globular. Wings closely packed. Carriage erect. Weights: Drake 5 lb (min.). Duck 4 lb. Colour: Orange-yellow bill. Eyes iris blue. Legs and feet orange-yellow. Plumage white.

Magpie: Origin British. Type: Head long and straight with lengthy neck. Body of great length with breadth and depth. Well-developed abdomen. Carriage somewhat racy, showing strength and great activity. Weights: Drake $5\frac{1}{2}$ to 7 lb; duck $4\frac{1}{2}$ to 6 lb. Colour black and white.

Penguin: Origin British. Type: Head straight with long and fairly thick neck. Inclined to favour the Pekin in shape of body with broad deep breast. Strong legs set well back. Carriage almost upright. Weights: Drake up to 9 lb; duck 7 lb. Colour black and white. Legs and webs dark.

GEESE

THE first poultry show of 1845 classified Common Geese, Asiatic or Knob Geese, and Any Other Variety. The first Book of Standards described the Toulouse and Embden. Peculiarly enough, these two breeds monopolised our Standards up to recent times, being the chief ones exhibited regularly at shows. At times other breeds have been exhibited, and now the Standards have been extended. The Greylag is said to be the ancestor of all our domestic geese, and the common goose of this country undoubtedly was the English Grey, although a white variety existed, and the Grey-back (Saddleback) may have come from an intercross.

The British Waterfowl Association classifies the following as Ornamental Geese: Canada, Egyptian and all breeds of British or foreign wild geese.

BRECON BUFF

AT different times attempts have been made to create a buff goose, and to Wales goes the credit of originating the Brecon Buff, founded on stock from Breconshire hill farms, the breed being recognised officially in 1934. Hardy, light in bone, with maximum flesh, and an active breed, it is also more prolific than the ultra-heavy breeds.

GENERAL CHARACTERISTICS: MALE AND FEMALE

Carriage: Upright and alert, indicative of activity.

Type: Body broad, well rounded and compact. Breast round, only slight indication of keel. Wings large and strong.

Head: Small and neat, no sign of coarseness. Eyes bright.

Neck: Long and thin, the throat showing no gullet.

Legs and Feet: Legs fairly short. Strong shanks. Straight toes connected by web.

Plumage: Hard and tight.

COLOUR

Plumage of both sexes a deep shade of buff throughout with markings similar to those of the Toulouse.   Ganders are usually not as deep coloured as geese. Bill pink. Eyes dark brown. Legs and webs pink or orange.

Standard Weights: Gander 19 lb (max.).   Goose 16 lb (max.).

SCALE OF POINTS

TYPE (BREAST 20, HEAD AND NECK 15, GENERAL CARRIAGE 10)	45
SIZE	15
LEGS AND FEET	5
COLOUR AND MARKINGS	25
CONDITION	10
	100

CHINESE

A PROLIFIC breed of the smaller-bodied type of goose is the Chinese, which has of recent years become popular on the show-bench. Its popularity has grown and it long since became Standardised in this country. It has at times been known as the Knob, Asiatic and Hongkong goose.

GENERAL CHARACTERISTICS: MALE AND FEMALE

Carriage: Upright, compact and active.

Type: Body compact and plump. Back reasonably short, broad, flat, and sloping, to give a characteristic upright carriage. Breast well rounded and plump, and carried high. Wings large, strong and high-up, carried closely. Stern well rounded, and well-developed paunch. Tail close and carried well out.

Head: Medium for size, and proportionate. Bill stout at base, symmetrical, medium for size.   Knob, large rounded and prominent (smaller in the female than the male). Eyes bold.

Neck: Long, swan-like, carried upright, but with graceful arch and refined (longer in the male than the female).

Legs and Feet: Legs reasonably short. Shanks strong, medium for length.   Toes straight, well spread and webbed.

336

Chinese Geese

Plumage: This should be reasonably tight.

COLOUR

White Variety: Plumage of both sexes pure white. Bill and knob orange. Shanks and toes orange-yellow. Eyes blue.

Brown (or Grey) Variety: Plumage of both sexes: Head dark russet-brown, with face fawn up to demarcation line above the eyes. Face band a definite white band or line from top of head to as far down the face as possible. Neck fawn with prominent dark russet-brown stripe down middle of back of neck and for its entire length. Back russet-brown. Breast greyish-fawn from under mandible well down to body where it becomes lighter. Thighs at side of breast russet-brown, each feather edged with a lighter shade of greyish-fawn, approaching white. Wing: bow and coverts medium russet-brown, each feather laced with a lighter greyish-fawn edging, approaching white; flights russet-brown. Stern, paunch and tail a lighter shade of greyish-fawn, approaching white, the tail having a broad band of russet-brown across with the light edging. Bill black or dark slate. Knob dark slate. Eyes brown. Shanks and toes orange.

Standard Weights: Gander 10 to 12 lb. Goose 8 to 10 lb.

SCALE OF POINTS

TYPE AND CARRIAGE	30
HEAD POINTS AND NECK	25
COLOUR	15
SIZE	10
CONDITION	10
LEGS AND FEET	10
	100

Disqualifications: Absence of knob and heavy gullet.

EMBDEN

As the Embden breed was also known originally as the Bremen one associates it with Germany although stock reached us from

Top, Embden (or White) and bottom, Brecon Buff Geese

Holland.   In Germany and North Holland, no doubt they crossed the Italian White with their native Whites, creating the Embden. When stock did reach this country our breeders crossed the birds with our own English Whites, and by careful selection increased the body-weight and quantity of meat, while standardising the breed for characteristics.

GENERAL CHARACTERISTICS: MALE AND FEMALE

Carriage: Upright and defiant.

Type: Body broad, thick and well rounded.   Back long and straight. Breast round with very little, if any, indication of keel.   Shoulders and stern broad. Paunch deep. Wings large and strong. Tail close and carried well out.

Head: Long and straight.   Bill fairly short and stout at the base. Eyes bold.

Neck: Long and swan-like, the throat uniform with the under-mandible and neck, i.e., without a gullet.

Legs and Feet: Legs fairly short.   Shanks large and strong.   Toes straight and connected by web.

Plumage: Hard and tight.

COLOUR

Plumage of both sexes pure glossy white.   Bill orange.   Eyes light blue.   Legs and feet bright orange.

Standard Weights: Gander 30 to 34 lb.   Goose 20 to 22 lb.

SCALE OF POINTS

TYPE (BREAST 20, HEAD 12, GENERAL CARRIAGE 12, NECK 10)	54
SIZE	20
COLOUR	10
CONDITION	10
LEGS AND FEET	6
	100

Serious Defects: Plumage other than white.   Any deformity.

ROMAN

Aɴᴏᴛʜᴇʀ of the smaller type of goose. The Roman was introduced into England from Italy about 1903, and there were other importations at later dates. Earliest arrivals often were marked with grey on the back, but were eliminated by selective matings for the pure White.

GENERAL CHARACTERISTICS: MALE AND FEMALE

Carriage: Active, alert, graceful, docile rather than defiant, with horizontal outline (particularly in female).

Type: Compact and plump, deep and broad, but reasonably long and well balanced. Back wide, flat and with gentle slope that is more pronounced in gander than goose. Breast full, well rounded, somewhat low and without prominence of keel. Wings large, long, strong, high-up and well tucked up to tail line. Stern well rounded off, paunch not too pronounced. Tail close, long and carried well out.

Head: Neat and well rounded (especially in female) symmetrical and refined. Face deep (particularly in female). Bill short and not coarse. Eyes bold, well up in skull.

Neck: Upright, medium length, refined (particularly in female) and without gullet.

Legs and Feet: Legs short, light boned, well apart. Toes straight and connected by web.

Handling: High proportion of meat to bone and offal. To be emphasised in judging.

Plumage: Sleek, short, tight and with glossy feathering.

COLOUR

Plumage of both sexes glossy white. Bill orange-pink. Eyes light blue. Legs and feet orange.

Standard Weights: Gander 12 to 14 lb. Goose 10 to 12 lb.

SCALE OF POINTS

TYPE AND CARRIAGE...................... 20

Continued on next page

Roman Scale of Points—continued

```
TABLE QUALITIES..............................  20
HEAD AND NECK...............................  15
COLOUR.......................................  15
IDEAL SIZE...................................  10
CONDITION....................................  10
LEGS AND FEET................................  10
                                             ——
                                            100
                                             ══
```

Serious Defects: Plumage other than white. Any deformity. Excessive weight or bone. Coarseness.

TOULOUSE

FRANCE originated the Toulouse and developed it for table purposes. Stock was sent over to England and our breeders crossed the birds with our own English Greys, and developed the breed for body-weight and quantity of flesh, as well as Standard characteristics of plumage colour, markings and type.

GENERAL CHARACTERISTICS: MALE AND FEMALE

Carriage: Thick set and somewhat horizontal, but not as upright in front as the Embden.

Type: Body long, broad and deep. Back slightly curved from the neck to the tail. Breast prominent, deep and full, the keel straight from stem to paunch, increasing in width to the stern and forming a straight underline. Shoulders broad. Wings large and strong. Tail somewhat short, carried high and well spread. Paunch and stern heavy and wide, with a full rising sweep to the tail.

Head: Strong and massive. Bill strong, fairly short and well set in a uniform sweep, or nearly so, from the point of the bill to the back of the skull. Eyes full.

Neck: Long and thick, the throat well gulleted.

Legs and Feet: Legs short. Shanks stout and strong-boned. Straight toes connected by web.

Plumage: Full and somewhat soft.

Toulouse Goose

COLOUR

Plumage of both sexes: Neck dark grey. Breast and keel rather light grey, shading dark to thighs. Back, wings and thighs dark steel-grey, each feather laced with an almost white edging, the flights without white. Stern, paunch and tail white, the tail with broad band of grey across the centre. Bill, legs and webs orange. Eyes dark brown or hazel.

Standard Weights: Gander 28 to 30 lb. Goose 20 to 22 lb.

SCALE OF POINTS

TYPE (HEAD AND THROAT 15, BREAST AND
  KEEL 10, TAIL, STERN AND PAUNCH 10,
  NECK 5, GENERAL CARRIAGE 15)........... 55
SIZE... 20
LEGS AND FEET............................. 5
COLOUR AND MARKINGS..................... 10
CONDITION................................. 10
                                           ———
                                           100
                                           ===

Serious Defects: Patches of black or white among the grey plumage. Slipped or cut wings. Any deformity.

OTHER BREEDS

Among other breeds of geese occasionally exhibited in this country are the following: African, one variety, brown (or grey) knobbed very similar to the Chinese, but generally twice its size and weight; Canadian, one variety, chiefly fawn and brown, edged with a dull white, with black bill, head, neck, rump, tail, legs and feet; Cereopsis, one variety, chiefly clear brown-grey, with dull white head, shoulders and upper part of back brown spotted, black tail, red legs, and black feet; English, two varieties, grey and white, very similar in colour to the Toulouse and Embden but generally half the size and weight of those breeds; and Sebastopol, one variety, pure white with frizzled and silky feathers on the back, wings and body, of extraordinary length and many of them trailing on the ground.

STANDARD FOR EGGS

The Poultry Club has authorised the following Standard and Scale of Points for judging eggs.

EXTERNAL

Shape: Showing ample breadth, good dome, with greater length than width, the top to be much roomier than the bottom and more curved. The bottom is more pointed in the hen egg than that of the pullet, but it should not be too pointed, and a circular, or even narrow shape is undesirable.

Size: Mere size is not a deciding point, without texture of shell combined. A pullet's normal egg when the bird starts to lay is $1\frac{3}{4}$ oz and increases quickly to 2 oz, exceeding that after several months of production. There is another increase in the hen egg after the moult.

Shell texture: Smooth, free from lines or bulges, evenly limed, smooth at each end, without roughness, porous parts or lime pimples.

Colour: Brown, tinted, white, cream, mottled, blue, green, olive, plum, etc.

Freshness, bloom and appearance: Shells to be clean, without dull or stale appearance; air-space to be small, as befits a new-laid egg, contents on candling to be free of blood or meat spots. Eggs may be washed in preparation.

Uniformity: Eggs forming a plate or exhibit, to be uniform in shape, shell grade, shell texture, size and colour.

INTERNAL

Yolk: Rich bright golden yellow, free from blood streaks or spots. Well rounded, smooth on surface, and well raised from albumen. One uniform shade. Blastoderm or germ spot not discoloured.

Albumen: Preferably white in colour, of dense substance, particularly around the yolk, which it raises. Outline of albumen to be seen. Free of blood spots.

Chalazæ: Each chalaza to resemble a thick cord of white albumen at each end of yolk, keeping it to centre of first or thickest albumen. Free of blood spots. Other layers of albumen less dense.

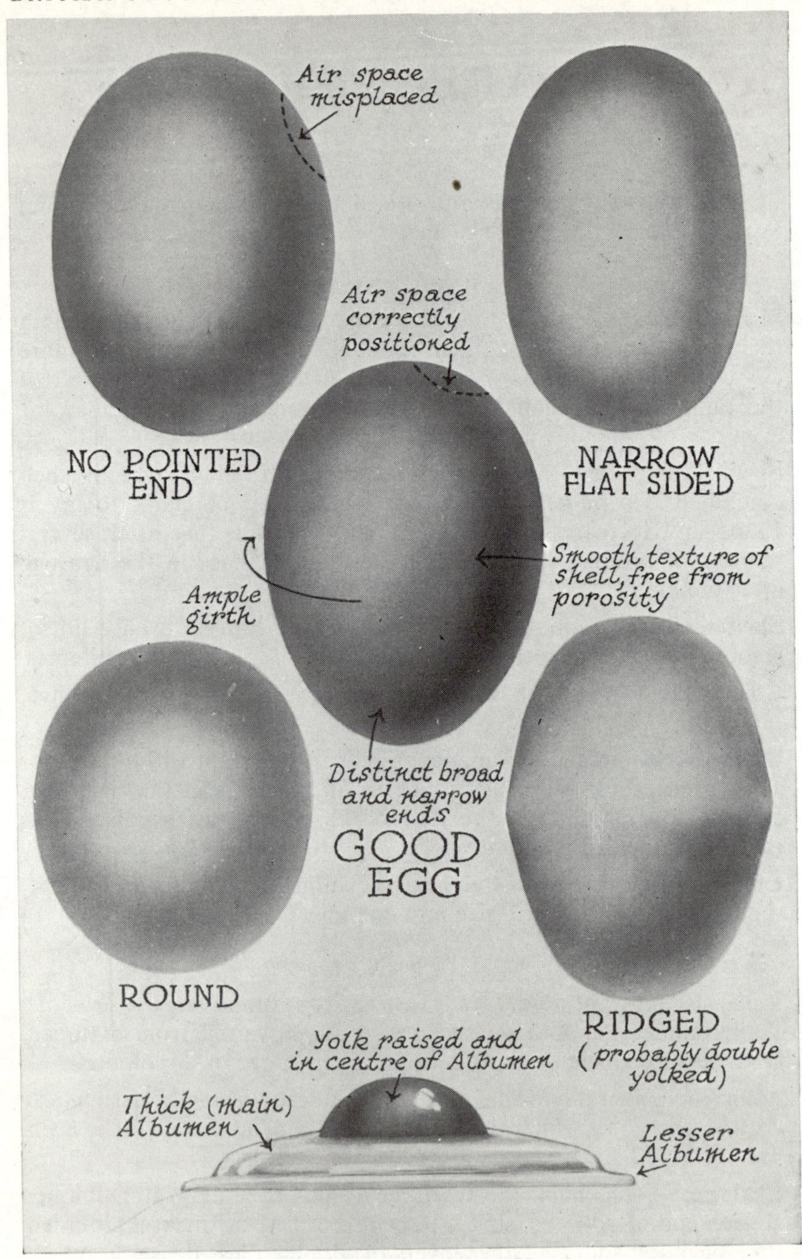

Diagrams show Standard and faulty eggs, and also the contents of a
Standard egg

Air-space: Very small, as befits a new-laid egg, the membrane adhering to shell.

Freshness: Indicated by small air-space, and unwrinkled top surface of yolk, and its height.   Stale yolks flop at edges.

SCALE OF POINTS

External		*Internal*	
SHAPE	20	YOLK	40
SIZE	20	ALBUMEN	30
SHELL TEXTURE	20	CHALAZÆ	10
COLOUR	20	AIR-SPACE	10
FRESHNESS, BLOOM, AND		FRESHNESS	10
APPEARANCE	20		
	100*		100

* May be maximum for each egg, or for a plate of eggs, whatever the number.   Add 5 points more for each egg for " matching and uniformity ".

Defects (for which eggs may be passed): More than one yolk. Staleness. Polished or over-prepared shells. Defective contents even when judged for externals. Addition of colouring to shells. Artificial polish or colouring would amount to disqualification and report to Poultry Club.

TABLE POULTRY

Type: Breast bone straight, long and deep. Back broad and flat. Thighs short and stout.

Quantity of Meat: Ample breast meat, to give plump, rounded appearance. Well fleshed thighs.

Quality of Meat: Softness of flesh with fineness and fine grain or texture of skin. Softness is highly essential.

Smallness of Bone and Absence of Offal: Fat to be firm and evenly distributed and must not be surplus in quantity. Fineness of bone desirable, in proportion to quantity and quality of meat.

Presentation and Colour: This includes killing, plucking and preparation, points being deducted for torn skin, dirty feet, discoloured flesh, water blisters on breast and similar blemishes in dead table birds. In live birds, it includes water bladders or hard fleshy growths on breast, discoloration of skin and blemishes.

SCALE OF POINTS

TYPE.. 20
QUANTITY OF MEAT..................... 20

Continued on next page

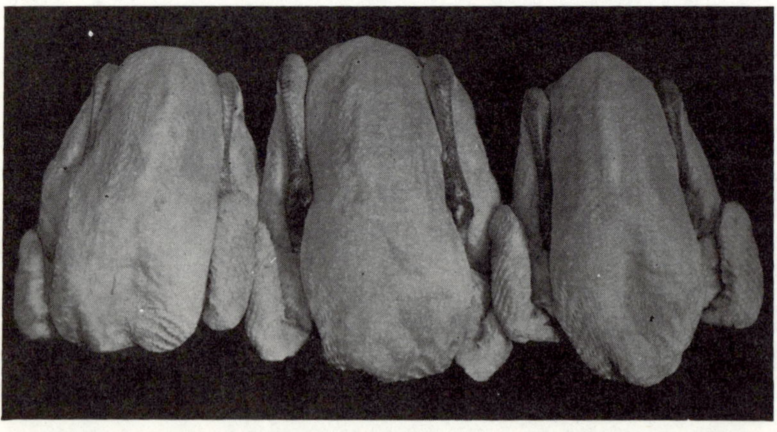

These carcasses typify most of the features necessary in good table poultry, such as sound formation of breast with well and evenly spread meat and fine quality bone

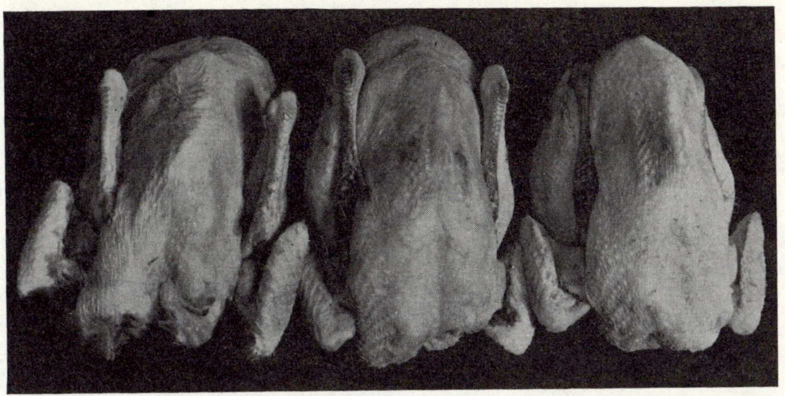

Compared with the photograph on the opposite page several typical faults are seen in this trio of table birds. Not only is matching uneven but the birds on either side have crooked breast bones while all show bruising or bad blemishes on the breast. The centre bird is a fair one, but the breast lacks plumpness and is seriously discoloured in parts as a result of faulty plucking

Table Poultry Points—continued

QUALITY OF MEAT.....................	20
SMALLNESS OF BONE AND ABSENCE OF OFFAL................................	20
PRESENTATION AND COLOUR............	20
	100

Additions:

PAIRS — MATCHING.....................	20
PACKS — MATCHING.....................	20
EFFICIENCY OF PACKING...............	20

Disqualifications: Broken breast bone. Artificial treatment to deceive the judge. Faking, over-age or over-weight (according to schedule stipulation). Not the breed or cross scheduled.

Serious Defects: Crooked breast bone. Hard growths or water cysts on breast.

Note.—Capons, surgical or chemical, should be shown in their own special classes.

DEAD TABLE DUCKLINGS

THE following Standard prepared by the British Waterfowl Association has been accepted by The Poultry Club:—

Preparation and Presentation: The correct method of exhibiting for market or table purposes is as follows:—The whole body, back and legs to be clean picked. The neck to be clean picked at least halfway up and finished evenly all round (the neck) without leaving too definite a ruff. The wings to be clean picked to the first joint from the body; all flights to be removed with the exception of the last three or four at the very end of the wing—these few remaining flights are required to help to hold the wing in place.

The wings should be twisted over (towards the body) so that the last joint comes underneath the first and second joints, when the flights will automatically come on the bird's back and point towards and nearly reach the tail of the duckling. The legs should then be twisted inwards and under, so that the feet also come on to the duckling's back, on top of the flights, thus holding the latter in place.

In a bird prepared in this way the feathers on the second wing joint will just show on each side of the breast, giving the duckling a finished and less naked appearance. Bill, feet and flights should have been washed and wiped clean, and necks dislocated.

No skewers, string, flour or singeing may be used and no decorative or distinctive marks are allowed. Exhibits should be free from tears, scratches, bruises, etc. Breaking of breast bone or any other trick will cause disqualification (the breaking of the pelvic bones is a normal method of preparation for market and is therefore permitted). When shown in couples, or packs of six, even matching as regards colour, shape and size is essential. A duckling may be of either sex for table purposes.

POINTS FOR JUDGING

Age: A duckling must not be more than 12 weeks of age or have passed its first feathering.

Condition: Meatiness; quality and fineness of meat and bone. Roundness and breadth of breast; the breast bone should not be noticeable if condition is really good. There should be no sharp angles, points or hollows around pelvis, tail or neck.   Absence of superfluous surface fat and offal gains points.

Colour: The whiter the better; yellowness, redness, blueness or discoloration loses points.   Where other points are equal, preference to be given to birds with flesh-coloured bills.

Size: Colour and condition being equal, preference should be given to size, providing texture of skin and fineness of bone are in proportion.

Presentation: Cleanliness of feet, legs and bill; absence of down feather on body and/or legs.

Matching: General uniformity in size, colour and condition to gain points, but each couple or pack to be judged for points as pair or pack and not separately.

SCALE OF POINTS

With a maximum of 100 points, the following are the maximum points for each of the stated qualifications:—

CONDITION	40
COLOUR	25
SIZE	20
PRESENTATION	15
	100

Disqualification: Any exhibit that has its breast bone broken or has suffered any other trick or faking or that is over-age.   One bird disqualified in a pair or pack class automatically disqualifies the whole exhibit.

GLOSSARY

Abdomen: Underpart of body from keel to vent.

A.O.C.: Any other colour.

A.O.V.: Any other variety.

Axial Feather: Small feather between wing primaries and secondaries.

Back: Top of body from base of neck to beginning of tail.

Bands: See " Pencilling ".

Bantam: Miniature fowl, formerly accepted as one-fifth the weight of the large breed it represented, but nowadays about one-fourth.

Barring: Alternate stripes of light and dark across a feather, most distinctly seen in the Barred Plymouth Rock.

Bay: A reddish-brown colour.   (See also " Wing Bay ".)

Beak: The two horny mandibles projecting from the front of the face.

Bean: A black spot or mark (generally raised) at the tip of the upper mandible of a duck's bill, seen in Cayugas and other breeds of waterfowl.

Beard: A bunch of feathers under the throat of some fowls, such as Faverolles, Houdans and some varieties of Polands. A tuft of coarse hair growing from the breast of an adult turkey male, also known as the " tassel ".

Beetle Brows: Heavy overhanging eyebrows, best seen in the Malay.

Bill: A duck's beak.

Blade: Rear part of a single comb.

Blocky: Heavy and square in build.

Booted: Feathers projecting from the shanks and toes, as in the Brahma, Cochin, and Booted Bantam.

Bow Legged: Greater distance between legs at the hocks than at knees and feet.

Brassiness: Yellowish foul colouring on plumage, usually on back and wing.

352

Breast: Front of a fowl's body from point of keel bone to base of the neck.   In dead birds, flesh on the keel bone.

Breed: A group of birds answering truly to the type, carriage and characteristics distinctive of the breed name they take.   There may be varieties within a breed, distinguished by differences of colour and markings.

Cap: A comb; also the back part of a fowl's skull.

Cape: Feathers under and at base of neck hackle, between the shoulders.

Capon: Strictly speaking a castrated male fowl, but term is also used to describe one treated chemically.

Carriage: The bearing, attitude, or style of a bird, especially when walking.

Caruncles: Fleshy protuberances on head and wattles of turkeys and Muscovy ducks.

Chicken: A term employed by the Poultry Club to describe a bird of the current season's breeding.

Cinnamon: A dark reddish-buff colour.

Cloudy: Indistinct (see " Mossy ").

Cobby: Short, round or compact in build.

Cock: A male bird after the first adult moult.

Cockerel: A male bird of the current year's breeding.

Cockerel-breeder: A term applied to birds, either male or female, selected to produce good standard-bred cockerels.

Collar: A white mark almost encircling the neck of the Rouen drake, also known as the " ring ".

Comb: Fleshy protuberance on top of a fowl's head, varying considerably in type and size and including cushion (Silkie), horn or V-shaped (La Flèche and Sultan), leaf or shell (Houdan), pea or triple (Brahma), rose (Hamburgh, Wyandotte, etc.), single (Cochin, Leghorn, etc.), cap (Redcap), cup (Sicilian Buttercup), strawberry or walnut (Malay), and raspberry (Orloff).

Concave Sweep: Hollow curve from shoulders to part way up the tail.

Condition: State of a bird's health, brightness of comb and face and freshness of plumage.

Coverts: Covering feathers on tail and wings.

Cow Hocks: Weakness at hocks (see " Knock-kneed ").

Crescent: Shaped like the first or last quarter of the moon.

Crest: A crown or tuft of feathers on the head; known also as " top knot " and in Old English Game as the " tassel ".

Crow Head: Head and beak narrow and shallow, like a crow.

Cuckoo Barring: Irregular barring where the two colours are somewhat indistinct and run into each other, as in the North Holland Blue, Cuckoo Leghorn and Marans.

Cup Comb: A comb somewhat resembling a tea-cup with the edges spiked as in the Sicilian Buttercup.

Cushion: A mass of feathers over the back of a female covering the root of her tail, and most prominently developed in the Cochin.

Cushion Comb: An almost circular cushion of flesh, with a number of small prominences over it, and having a slight furrow transversely across the middle, as in the Silkie.

Daw Eyed: Having pearl coloured eyes.

Deaf-ears: See Ear-lobes.

Dewlap: The gullet (so called), seen to the best advantage in adult Toulouse geese. Loose pouch of skin on throat under the beak.

Diamond: The wing bay. A term commonly used among Game fanciers.

Dished Bill: Depression or hollow in the upper line of the bill of a duck or drake.

Dished Lobe: Lobe that is hollow in the centre.

Double Comb: See " Rose Comb ".

Double Laced: Two lacings of black as on an Indian Game female's feather. First there is the outer black lacing round the edge of the feather and next the inner or " second " lacing (see "Lacing").

Down: Initial hairy covering of baby chicks, ducklings, etc. Also the fluffy part of the feather below the web and small tufts sometimes seen as faults on toes and legs of clean legged breeds (see " Fluff ").

Drake: Male duck.

TYPES OF COMB: 1—Rose, leader following line of neck. 2—Triple or pea. 3— Rose, short leader. 4—Walnut. 5—Cap. 6—Mulberry. 7—Medium single. 8—Large single. 9—Cup. 10—Rose with long leader. 11—Leaf. 12—Horn. 13—Small single. 14—Folded single. 15—Semi-erect single

Key opposite

front of the fowl's head. It has also been referred to as resembling two escallop shells joined near the base, the join covered with a piece of coral. Seen to the best advantage on a Houdan male.

Leg: The shank or scaly part.

Leg Feathers: Feathers projecting from the outer sides of the shanks of such breeds as Brahmas, Cochins, Faverolles, Langshans and Silkies.

Lesser Sickles: See " Sickles ".

Lobes: See " Ear-lobes ".

Lopped Comb: Falling over to one side of the head.

Lustre: See " Sheen ".

Main Tail Feathers: See " Tail Feathers ".

Mandibles: Horny upper and lower parts of beak or bill.

Marking: The barring, lacing, pencilling, spangling, etc., of the plumage.

Mealy: Stippled with a lighter shade, as though dusted with meal, a defect in buff coloured fowls.

Moons: Round spangles on tips of feathers.

Mossy: Confused or indistinct marking; smudging or peppering. A defect in most breeds.

Mottled: Marked with tips or spots of different colour.

Muff: Tufts of feathers on each side of the face and attached to the beard, seen in such breeds as Faverolles, Houdans, and some varieties of Poland; also known as " whiskers ".

Muffling: The beard and whiskers, i.e., the whole of the face feathering except the crest. In Old English Game the Muffed variety has a thick muff or growth of feathers under the throat, differing in formation from that of the breeds named under " Muff ".

Mulberry: See " Gypsy Face ".

Open Barring: Where the bars on a feather are wide apart.

Open Lacing: Narrow outer lacing, which gives the feather a larger open centre of ground colour.

Outer Lacing: Lacing around the outer edge of a feather as opposed to " inner " lacing.

Parti-coloured: Breed or variety having feathers of two or more colours, or shades of colour.

360

Pea Comb: A triple comb, resembling three very small single combs joined together at the base and rear, but distinctly divided, the middle one being the highest; best seen on the head of a well bred Brahma.

Pearl Eyed: Eyes pearl coloured. Sometimes referred to as " daw eyed ".

Pencilled Spikes: The spikes of a single comb that are very long and narrow; little broader at the base than at the top; generally a defect.

Pencilling: Small markings or stripes on a feather, straight across in Hamburgh females (and often known as bands); or concentric in form, following the outline of the feather, as in Brahmas (Dark), Cochins (Partridge), Dorkings (Silver Grey), and Wyandotte (Partridge and Silver Pencilled) females, and fine stippled markings on females of Old English Game and Brown Leghorns.

Peppering: The effect of sprinkling a darker colour over one of a lighter shade.

Primaries: See " Flights ".

Primary Coverts: See " Flight Coverts ".

Pullet: A female fowl of the current season's breeding.

Pullet-breeder: A term applied to birds, either male or female, selected to produce good standard-bred pullets.

Pupil: Black centre of eye.

Quill: Hollow stem of feathers attaching them to the body.

Raspberry Comb: A comb somewhat resembling a raspberry cut through its axis (lengthwise) and covered with small protuberances.

Reachy: Tall and upright carriage and " lift " as in Modern Game.

Ring: See " Collar ".

Roach Back: Humped back.

Rose Comb: A broad comb, nearly flat on top, covered with small regular points or " work ", and finishing with a spike or leader. It varies in length, width, and carriage according to breed.

Rust: A patch of red-brown colour on the wings of the females of some breeds, chiefly those of the black-red colour; brown or red marking in black fluff or breast feathers; known also as " foxiness " in females.

Saddle: The posterior part of the back, reaching to the tail of the male, and corresponding to the cushion in a female.

Saddle Hackle: See " Hackles ".

Sandiness: Giving the appearance of having been sprinkled with sand.

Sappiness: A yellow tinge in plumage.

Secondaries: The quill feathers of the wings which are visible when the wings are closed.

Self Colour: A uniform colour, unmixed with any other.

Serrations: " Saw tooth " sections of a single comb.

Shaft: The stem or quill part of the feather.

Shafty: Lighter coloured on the stem than on the webbing; a desirable marking in Dark Dorking females and Welsummers. Generally a defect in other breeds.

Shank: See " Leg ".

Shank Feathering. See " Feather Legged ".

Sheen: Bright surface gloss on black plumage. In other colours usually described as lustre.

Shell Comb: See " Leaf Comb ".

Shoulder: The upper part of the wing nearest the neck feather. Prominent in Game breeds where it is often called the shoulder butt. (See also " Wing Butt ".)

Sickles: The long curved feathers of a male's tail, usually applied to the top pair only (the others often being called the " lesser " sickles), but sometimes used for the tail coverts.

Side Sprig: An extra spike growing out of the side of a single comb.

Single Comb: A comb which, when viewed from the front, is narrow, and having spikes in line behind each other. It consists of a blade surmounted by spikes, the lower (solid) portion being the blade, and the spaces between the spikes the serrations. It differs in size, shape, and number of serrations according to breeds.

Slipped Wing: A wing in which the primary flight feathers hang below the secondaries when the wing is closed. This condition is often allied with split wing, in which primaries and secondaries show a very distinct segregation in many breeds of bantams.

LEG TYPES: 1—Clean legged, flat sided (Leghorns). 2—Clean legged round shanks (Game). 3—Heavy feather legged, and feathered toes. *i.e.* foot feather. 4—Feather legged, no feathers middle toe (Croad Langshan). 5—Short round shanks (Indian Game) 6—Five toed (Dorking). 7—Slightly feathered shanks (Modern Langshan). 8—Feather legged and vulture hocked. 9—Thin round shanks (Modern Game). 10—Mottled shanks (Ancona). 11—Mottled and five toed (Houdan). 12—Feather legged and five toed (Faverolle)

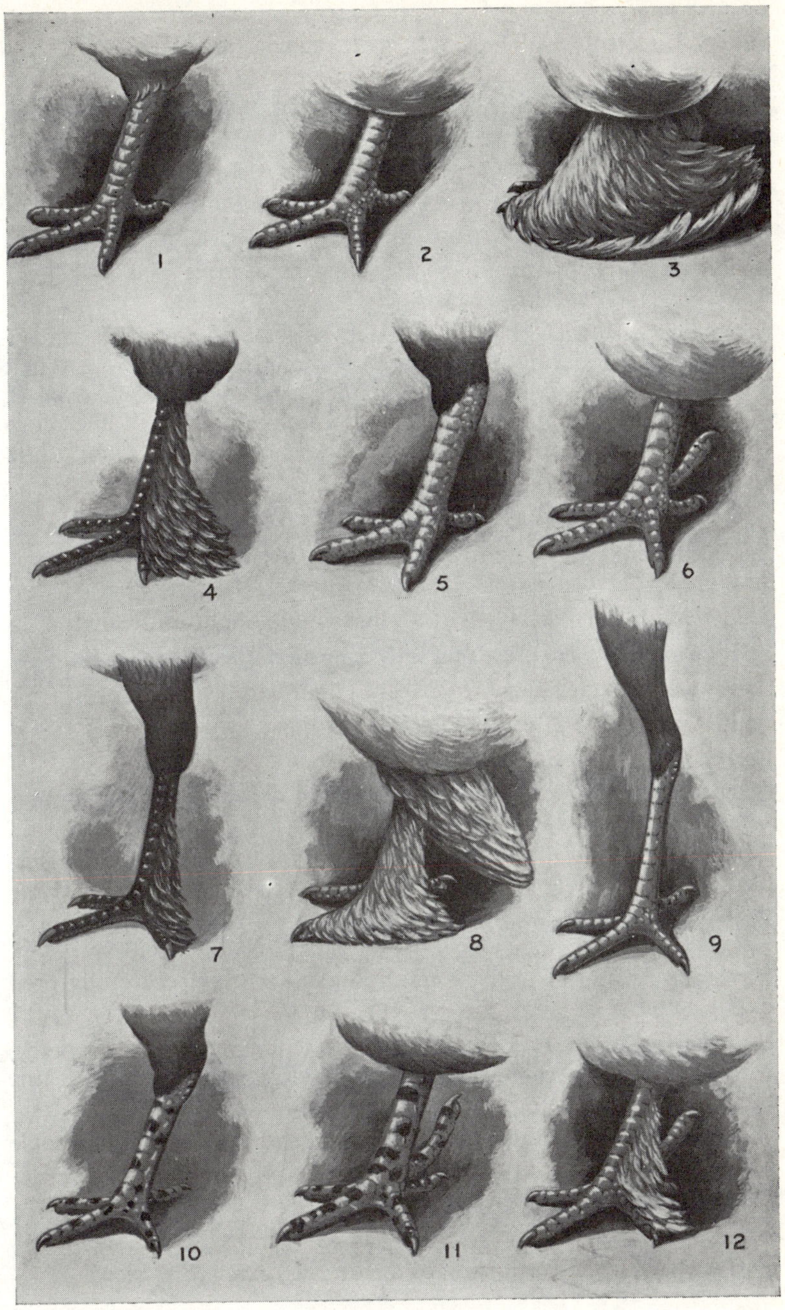

Smoky Undercolour: Defective grey pigment in the undercolour of a bird.

Smut: Dark or smutty colour where undesirable, such as in undercolour.

Soft Feathered: Applied to breeds other than the hard feather group of Indian and Jubilee Game, O.E. Game, Aseel, Malay, and Modern Game.

Sootiness: Grey or smokiness creeping in where it is not wanted, usually in undercolour.

Spangling: The marking produced by a spot of colour at end of each feather differing from that of the ground colour. When applied to a laced breed, as the Poland, it means broader lacing at the tip of each feather. The spangle of circular form is the more correct, since when of crescent or horse shoe shape it favours the laced character.

Spike: The rear leader on a rose comb.

Splashed Feather: A contrasting colour irregularly splashed on a feather.

Split Comb: The rear blade of a single comb is split or divided.

Split Crest: Divided crest that falls over on both sides.

Split Tail: Decided gap in middle of tail at base.

Split Wing: See " Slipped Wing ".

Sprig: See " Side Sprig ".

Spur: A projection of horny substance on the shanks of males, and sometimes on females.

Squirrel Tail: A tail, any part of which projects in front of a perpendicular line over the back; a tail that bends sharply over the back and touches, or almost touches, the head, like that of a squirrel.

Strain: A family of birds from any breed or variety carefully bred over a number of years.

Strawberry Comb: A comb somewhat resembling half a strawberry, with the round part of the fruit uppermost; known also as the walnut comb.

Striping: The very important markings down the middle of hackle feathers, particularly in males of the Partridge variety.

Stub: Short, partly grown feather.

Sub-variety: See " Variety ".

Surface Colour: That portion of the feathers exposed to view.

364

Sword Feathered: Having sickles only slightly curved, or scimitar shaped, as in Japanese bantams.

Symmetry: Perfection of outline, proportion; harmony of all parts.

Tail Coverts: See " Coverts ".

Tail Feathers: Straight and stiff feathers of the tail only. The top pair are sometimes slightly curved, but they are generally straight or nearly so. In the male fowl, main tail feathers are contained inside the sickles and coverts.

Tassel: See " Crest " and " Beard ".

Thigh: That part of the leg above the shank, and covered with feathers.

Thumb-marked Comb: A single comb possessing indentations in the blade; a defect.

Ticked: Plumage tipped with a different colour, usually applied to V-shaped markings as in Anconas. Also small coloured specks on any part of feathers of different colour from that of the ground colour.

Tipping: End of feathers tipped with a different coloured marking.

Top Colour: See " Surface Colour ".

Top Knot: See " Crest ".

Tri-coloured: Of three colours. The term refers chiefly to buff and red fowls, and is generally applied only to males when their hackles and tails are dark compared with the general plumage, and the wing bows are darker; a fault.

Trio: A male and two females.

Triple Comb: See "Pea Comb".

Twisted Comb: A faulty shaped pea or single comb.

Twisted Feather: The shaft and web of the feather are twisted out of shape.

Type: Mould or shape. (See " Symmetry ".)

Undercolour: Colour seen when a bird is handled—that is, when the feathers are lifted; colour of fluff of feathers.

V-shaped Comb: See " Horn Comb ".

Variety: A definite branch of a breed known by its distinctive colour or marking—for example, the Black is a variety of the Leghorn. Sub-variety, a sub-division of an established variety, differing in shape of comb from the original—for example, the rose-combed Black is a sub-variety of the Black Leghorn. Thus the breed includes all the varieties and sub-varieties which would conform to the same standard type.

Vulture Hocks: Stiff projecting quill feathers at the hock joint, growing on the thighs and extending backwards.

Walnut Comb: See " Strawberry Comb ".

Wattles: The fleshy appendages at each side of base of beak, more strongly developed in male birds.

Web: A flat and thin structure. Web of feather: the flat or plume portion. Web of feet: the flat skin between the toes. Web of wing: the triangular skin seen when the wing is extended.

Whiskers: Feathers growing from the sides of the face (see " Beard " and " Muff ").

Wing Bar: Any line of dark colour across the middle of the wing, caused by the colour or marking of the feathers known as the lower wing coverts.

Wing Bay: The triangular part of the folded wing between the wing bar and the point (see " Diamond ").

Wing Bow: The upper or shoulder part of the wing.

Wing Butt or Wing Point: The end of the primaries; the corners or ends of the wing. The upper ends are more properly called the shoulder butts and are thus termed by Game fanciers. The lower, similarly, are often called the lower butts.

Wing Coverts: The feathers covering the roots of the secondary quills.

Work: The small spikes or working on top of a rose comb.

Wry Back: A distorted bone structure usually causing a hump-backed condition.

Wry Tail: A tail carried awry, to the right or left side of the continuation of the backbone, and not straight with the body of the fowl.

SITTERS AND NON-SITTERS

GENERALLY speaking these divide themselves—Heavy breeds being sitters and the Light breeds non-sitters—the former comprising mainly American and Asiatic breeds with Indian Game, Sussex, and Dorkings of British origin, while the latter are generally of Mediterranean origin. Unlike certain other countries, however, which classify three categories—Light, Medium and Heavy—Great Britain adheres to two classes only. There are, therefore, certain exceptions to the foregoing generalisation. The breeds are divided as follows:—

Sitters: Aseel, Australorp, Barnevelder, Brahma, Brockbar, Brussbar, Cochin, Crève-Cœur, Croad Langshan, Dorbar, Dorking, Essex, Faverolle, Frizzle, Houdan, Indian Game, Ixworth, Jersey Giant, Jubilee Indian Game, La Flèche, Malay, Marans, Malines, Marsh Daisy, Modern Game, Modern Langshan, New Hampshire Red, Norfolk Grey, North Holland Blue, Old English Game, Orloff, Orpington, Phœnix, Plymouth Rock, Rhodebar, Rhode Island Red, Scots Dumpy, Silkie, Sultan, Sumatra Game, Sussex, Wherwell, White Surrey, Wyandotte, Yokohama.

Non-Sitters: Ancona, Andalusian, Bresse, Cambar, Campine, Coveney White, Hamburgh, Lakenvelder, Legbar, Leghorn, Minorca, Old English Pheasant Fowl, Poland, Redcap, Scots Grey, Sicilian Buttercup, Sicilian Flower Bird, Spanish, Welbar, Welsummer, Wyndham Black.

Bantams take the same classification, i.e. Heavy or Light, as their large prototypes.

In ducks the Indian Runner is probably the only real light breed, Campbells and Orpingtons being regarded as mediums and Aylesburies and Pekins as heavies.

There are no divisions in Geese and Turkeys. Trend in the latter is, however, to develop smaller carcasses with broader breasts, hence the growing demand for breeds like Small Whites which give eight to ten lb birds at reasonably early ages.

DEFECTS AND DEFORMITIES

THE Judges' Administration Board of The Poultry Club instructs its judges to work continuously for stock improvement in making their awards. They should keep in mind, therefore, the suitability of exhibits for the breeding pen, penalising those defects that affect reproductive values or detract from what may be regarded as the highest merits of such birds.

Uniformity in judging and exhibiting is sought, the aim being to make show pens reliable " shop windows ", displaying birds that buyers may claim with every confidence in their soundness and reproductive ability.

TO BE PASSED OR PENALISED—The following are given as deformities and defects for which judges must penalise or pass an exhibit according to the seriousness of the defect.

Head Points: Crossed or deformed beak. Malformation of beak. Badly dished bill in ducks. Blindness. Defective eyesight. Defective pupils. Odd eye colour. Comb that closes the nostrils.

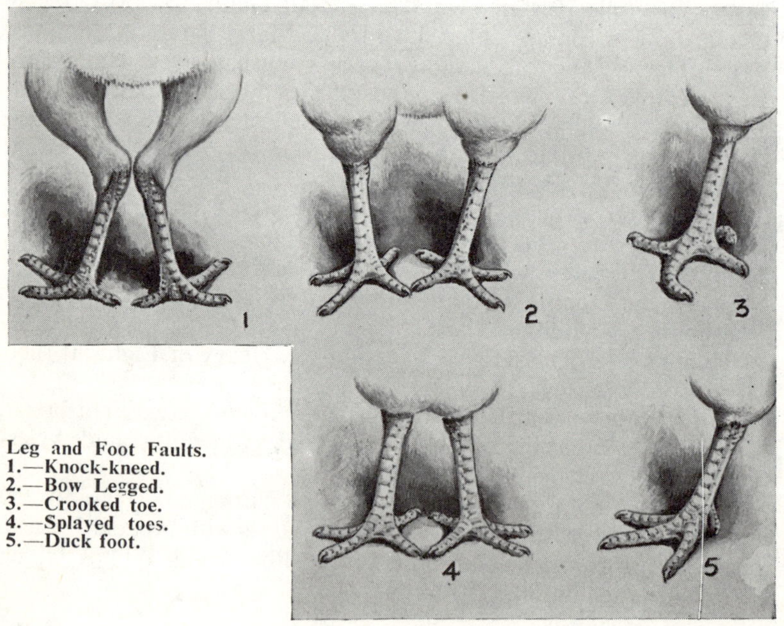

Leg and Foot Faults.
1.—Knock-kneed.
2.—Bow Legged.
3.—Crooked toe.
4.—Splayed toes.
5.—Duck foot.

Left—Ingrown leader
Right—Short of leader, and uneven wattles

Left—Rose comb falling to side and blocking vision
Right — Bad leader and coarse worked comb

Left—Beefy, and with part of blade too far forward
Right—Badly curved at rear end with spikes falling over

Left—Thumb mark and side sprig
Right—Fly-away comb

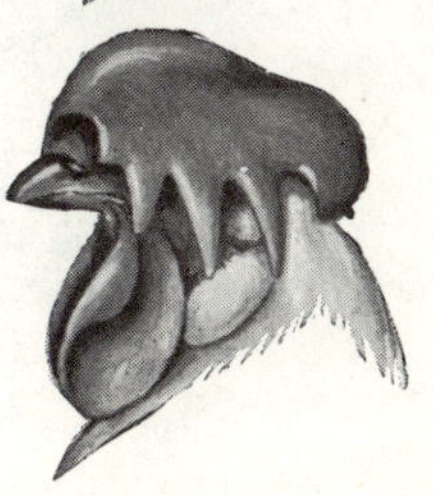

Left—Double folded comb
Right—Flop comb blocking vision

COMB FAULTS

Side sprigs or double end on a single comb. Excessive fall of rose-comb on a single base. Split combs at blade. Fall-over comb that obstructs vision. Malformed combs. Defective serrations. Peculiar head carriage. White in face. Wry neck. Indications of brain or nerve affection. Badly distended and sagging crop.

Dished bill

Back: Any deformity. Rounded or curved spine. Weak back formation. One bone higher than the other giving the back a lop-sided appearance.

Tail: Wry. Squirrel. Defective parson's nose. Split or divided tail feathers or badly twisted feathers in tail. High tail in excess.

Bone Structure: Pigeon breast. Seriously deformed breast bone. Malformation of breast bone that interferes with the internal organs. Down behind and curved end of breast bone which leads to drooping abdomen. Dented breast bone from perching. Enlargement on breast bone of turkey. Broken or malformed pelvic bones. Faulty stance.

Wings: Badly twisted or curled wing feathers. Slipped or drooping wing. Split wing, in a serious form, with large gap between primaries and secondaries. Defective wing formation in waterfowl.

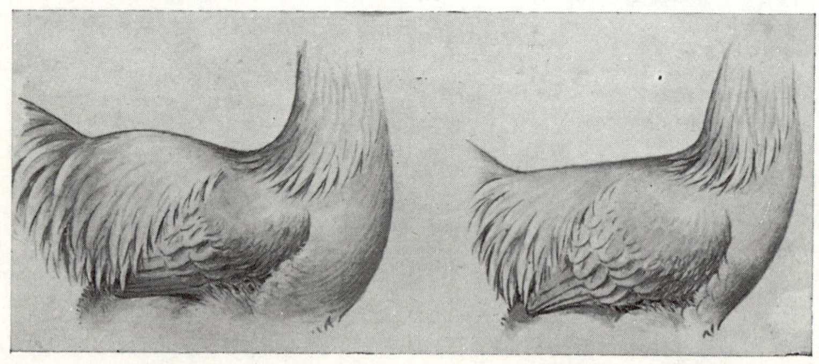
Roach back   Cut-away breast

370